BAR CODES

Design, Printing and Quality Control
William H. Erdei

McGraw-Hill, Inc.

New York St. Louis San Francisco Auckland Bogotá Caracas
Lisbon London Madrid México Milan Montreal New Delhi Paris
San Juan São Paulo Singapore Sydney Tokyo Toronto

BAR CODES
Design, Printing & Quality Control
by William H. Erdei

Copyright © 1993, by McGRAW-HILL INTERAMERICANA DE MEXICO, S.A. DE C.V.
 Atlacomulco 499-501, Fracc. Ind. San Andrés Atoto,
 53500 Naucalpan de Juárez, Edo. de México

ISBN 0-07-019448-3 English Language Edition

Trademarks: All product names are trademarks of their respective companies

1234567890   9087654123

Impreso en México   Printed in Mexico

To my children Florence, Andrea, Brenda and Alan
To Irene

TABLE OF CONTENTS

Preface

The second half of the twentieth century saw the development of a new technology called electronics. This new technology was amazing in its particular way of gathering knowledge from such diverse areas as quantum physics, solid state chemistry, semiconductors, micro systems and digital techniques, and was even more impressive because of its immediate applications. The greatest expression of this new technology, and an unmistakable symbol of the times, is the computer.

Humans have created these electronic brains capable of storing and processing huge amounts of data in their own binary language. Now we find the first approach by this system to human language, bar codes. This new electronic identification system is the latest way to symbolize different kinds of data directly into a binary system. Especially created to be scanned and processed by computers, bar codes maximize interaction possibilities with human beings and enhance the utilization of available technology.

Bar codes are a reality, successfully implemented, tested, and used every day in more than 200,000 installations all over the world, each with its branches and thousands of checkout counters or points of sale. To meet different needs, today we have several kinds of codes, a variety of computers within a wide price range, features, software and technology available for hundreds of applications. The next step is for us to put the potential of this mighty technical and commercial tool into practice, through the proper understanding and development of its advantages.

The whole world is changing its commercial patterns; this process demands assimilation of modern technology capable of producing quick results in boosting production and commercial exchange, both in domestic markets and foreign trade. Computers use the bar code language all over the world. In order to reach them, we must use their same language. This book is aimed at all individuals, companies and organizations who, facing the possibility of adopting bar codes, are wondering:

- What exactly is this, what can I gain from it?
- Where do I place it, and how do I print it?
- How do I set it up, operate and checkit?
- Who can help me and how?

The theoretical and practical procedures of each step in bar code implementation plus the different packaging and printing technologies and possible bar code combinations make this process so long and complex that it is almost impossible to include them all in a single book.

The purpose of this book is to discuss the basics of different bar codes for commercial use, their specifications, packaging code design, printing techniques and quality control, with the aim of producing a new and permanent relationship between packaging designers, printers, packaging industries, manufacturers and distributors that can result in the choice of a satisfactory, workable system.

The intention of this book is to provide readers with the necessary tools to discover, master, and benefit from what this new language technology offers us today; a language that is rapidly becoming a leader in international trade.

William H. Erdei

Acknowledgements

I would like to express my appreciation to those who contributed to the publication of this book and the following organizations and persons who authorized me to publish their literature:
- Automatic Identification Manufacturers (AIM), United States
- Coras Group, United States, Argentina, Brazil, Uruguay
- Ergi GmbH, Germany
- CC1, Inc., United States
- Eng. Frederick Braun, D. Passeron and D. Barmat, Argentina
- Hijos de Mariano Blasi SA, Spain
- International Article Numbering Association (EAN), Belgium
- Intermec Corp., United States
- Morton International, United States
- Photographic Sciences Corp., United States
- RJS Inc., United States
- Schiavi Cesare C.M. Spa, Italy
- Stork Screens, Holland; Stork Graphics, Spain
- Symbol Technologies Inc., United States
- Uniform Code Council Inc. (UCC), United States

Introduction

This book was designed to offer the reader both the general and the specific points of view on bar-coding technology and language and was compiled to assist all areas affected by commercial bar-coding applications for domestic and international markets.

As this is a technical book, the reader need not read it through from beginning to end. Instead readers who do not wish to read the whole book should go directly to the main specific-interest chapters (M) or the general interest chapters (G) as suggested below.

- MANUFACTURERS: Of all bar coded retail items to be sold at supermarkets or other points of sale

Packaging Design Department	M: 6,7,9	G:	2 to 5
Production Management	M: 6,8,9	G:	2 to 5,10
Quality Control Lab.	M: 8,9	G:	2 to 6
Export-Import Department	M: 2 to 5	G:	9,10
Labeling and Traffic areas	M: 5,6,9,10	G:	2 to 4
Sales Department	M: 2,5,6	G:	9
Bar Code Responsible	M: 1 to 10		

- PACKAGING INDUSTRIES: Graphic arts, flexible packaging converters, rigid packaging industries, all types of printers, books and magazine's editors.

Sales Department	M: 2,6	G:	7,8,10
Production Department	M: 6,7	G:	2,3,4,8
Quality Control Department	M: 6,8	G:	3,4
Bar Code Responsible	M: 1 to 10		

- DISTRIBUTORS: Supermarkets, drugstores and all retail stores

Project Department	M: 2,5,10	G:	1,3 to 8
Quality Control Department	M: 8	G:	2 to 6,9
Labeling and Traffic Department	M: 5,9	G:	2 to 4,10
Bar Code Responsible	M: 1 to 10		

- SERVICE SUPPLIERS
  - Labels, Mailing Services   M: 6,7,9   G: 2 to 5
  - Marketing and Commercialization   M: 2,5,6   G: 3,4,9
  - Industrial Packaging Designers   M: 2,6   G: 3 to 5,7,8,9
  - Import-Export Trading Companies   M: 2 to 6   G: 9,10
  - Consultants   M: 1 to 10

- HIGH SCHOOLS, UNIVERSITIES, TECHNICAL SCHOOLS
  - Industrial Packaging Design, Applied Arts   M: 1,6,7   G: 2 to 5,9
  - Packaging   M: 1,6,8   G: 2 to 5,9
  - Graphic Arts   M: 1,6,7,8   G: 2 to 4,9

- TECHNICAL INSTITUTES, ASSOCIATIONS
  - Packaging   M: 6   G: 2 to 4,7,8
  - Printing, Graphic Arts, Converting   M: 6,7,8   G: 2 to 5
  - Labels, Tags, ID   M: 5,6,9,10   G: 2 to 4,8
  - Exporters and Importers   M: 1,2,5,9   G: 6,10
  - Books, Magazines, Videotapes   M: 4,6   G: 1,2,9

Automatic
Identification
Systems

Identification systems are currently applied to individuals and objects in the form of magnetic, optical, acoustical or printed records.

These records usually consist of a coded element bearing data and a scanning device capable of recognizing the data. Information is then fed into a computer, where the identification is decoded, verified, matched and accepted in order to make a logical decision such as the identification of persons for access to bank accounts, restricted areas, computers, telephone lines, corporate premises, their homes, remote control units and credit cards, among others.

Modern systems are automatic. Automation accelerates processes, prevents human error and makes whole operations more reliable and effective.

These same systems are used for object identification, specifically when related to commercial activities; the bigger the marketing level, the more necessary exact identification of an item becomes, to make manufacturers, retailers, distributors and consumers aware of an item's characteristics, origin, location and destination, cost and sale price, verification and control, accounting and administration, statistics and inventory levels. Some applications of these systems include:

A. Electronic Vision

Reading is handled by TV cameras and/or sets of mechanical or photoelectric cells connected to computers programmed to distinguish beetween shapes, images, colors and products for quality control, positioning, inspection and security. Industrial robots generally use this identification system, which is found mainly in the automobile and electronics industries.

Figure 1-1

B. Magnetic Stripe

Electromagnetic data is recorded on tape sections, generally attached to plastic films such as credit cards, ID cards, payment services and miscellaneous items such as cards for highway tolls and transportation in some countries.

The magnetic tape is basically similar to standard audio cassette tape but tailored to a rigid base film. Sometimes the magnetic coating is applied directly on to the plastic base, as with credit cards. When the magnetic tape is scanned, data is recognized, decoded and processed.

Figure 1-2 (*Courtesy of VISA*).

C. Magnetic Character Recognition

Characters hold data in their own format, pattern and thickness and are mechanically or magnetically scanned and recognized. These are usually numerical characters, allowing data to be immediately understood by people. This is the standard way to encode data on checks and other common commercial instruments, punched cards, currency notes (in some countries) and mechanized mailing systems. Checks usually bear a series of numbers for automatic processing in their lower section, based on this magnetic scanning system, called MICR (magnetic ink character recognition).

In Europe, code CMC7 is used, where each digit is formed by seven small vertical bars and spaces forming a binary system of 1s and 0s. According to their location, they define numbers from 0 to 9 as well as some symbol-encoded data such as: bank name, branch office, account number and check number (this system is an early predecessor of bar codes).

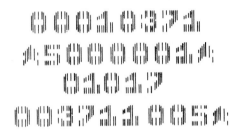 

Figure 1-3 *CMC7 magnetic character code used on European and Latin American checks.*

In the United States, code E13B is used for the same purpose. Bank and check data are magnetically encoded in vertical 13/1000-inch wide lines that form the numerical characters we can read.

Figure 1-4 *E13B magnetic character code used on U.S. checks.*

3

D. Optical Character Recognition (OCR)

These are also printed characters whose shape encodes the data to be processed. They are scanned by a light beam and decoded by mathematical algorithms to digital, analogical or ASCII formats.

The scanning beam can be either fixed or mobile, visible or not (infrared). The light source can be either color (incandescent) or coherent (laser), solid state (LED) or gas (helium-neon). These systems are being replaced by Bar Codes for a broad range of commercial applications.

Assigning a code and labeling each item (as in old supermarkets) is not an automatic identification system, since each item's marking and reading processes are done manually, whereas OCR refers only to automatic systems.

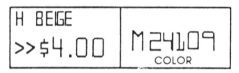

Figure 1-5

E. Voice Data Entry

This is a relatively new technology consisting of a computer programmed to recognize and understand words from a given glossary and transform this data into logic instructions. The system is also capable of uttering words in a synthesized voice.

Operators have microphones and earphones that allow them to talk and listen to the computer. The system is most suitable for those jobs requiring the operator's eyes and hands to be busy, as with critical lab activities, electronics, inventories, or material processing. Most systems are actually fed with the operator's voice. It is hoped they will soon be capable of recognizing any human voice.

Figure 1-6 *(Courtesy of AIM).*

F. Radio Frequency and Infrared

These are systems for simultaneous transmission and identification, as identity data are encoded and decoded in different ways that, after being recognized, allow active or passive access to computer

4

memory and commands. The system applies to aggressive environments, dangerous chemicals or high temperature material handling, industrial process control and industry identification where the action occurs far away from where decisions are made.

Many household applications such as remote controls for TV sets, VCR's and toys are available today. As a rule, equal RC units will activate equal models of TVs or receivers. Remote-control-activated electronic garage door locks and some car' RC units require complex encoded data to make them safer from tampering. Old ultrasonic remote control units are virtually no longer used.

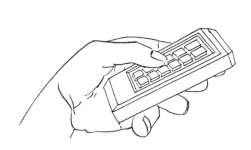 

Figure 1-7

G. Biometric Identification

Biometric ID systems recognize certain physiological characteristics that are different in every human being, such as fingerprints, voice patterns, blood vessel arrangements in the retina, wrist or hand, signatures and hand writing patterns. Personnel identification for access control and high security are the main applications of this new ID system mainly oriented to control access to government, military and other security areas. The best known application of biometric ID system is the fingerprint scanner that stores fingerprints digitally; the person places his or her finger on a reader that will verify whether it matches the stored data thus allowing or denying the access requested.

H. Electronic Data Interchange (EDI)

This new system of inter-company communication is rapidly becoming very popular as it is mainly based on both computers and telephone lines available in most companies all over the world. EDI logs supplier and customer computers allowing electronic exchange of common standard commercial data on orders and documents such as purchase orders, invoices and shipping information. Data is entered into the system as usual by keyboard or any auto ID system such as bar codes, which are very often used both as input and output links between EDI goods and processes being electronically connected. EDI networks replace mail or courier delivery of commercial documents between companies, reducing delays and errors and increasing data speed, quality and profits.

I. Smart Card Identification

Smart Cards and optical cards are credit-card-sized auto ID active systems containing miniature solid-state electronic devices such as microchips able to store digital data in their memory system. They can be programmable or not. Smart cards and Optical DRAW (directly read after writing) cards are alterable, allowing data only to enter but not permitting stored data modification. Unalterable smart cards are ROM based (read only memory). Optical card data storage looks like optical disks but acts on parallel rather than circular tracks.

These new auto ID systems are very small, ligth and portable, although they allow a lot of hidden data storage, but they are fragile, and data is hard to modify. These devices permit other standard auto ID systems on their exterior, such as bar codes and magnetic stripes as well as standard text.

J. Bar Codes

No doubt this is the best known and most widely adopted identification system especially for automatic processing of mass consumption items and is the main subject of this book.

★ C O D E 3 9 ★

A 7 1 4 5 B

8 7 3 0 2 1 5 7

0 73568 41259 2

4 025678 923563

Figure 1-8

2

Bar Code
Systems

Bar code, the most advanced automatic identification technology available applicable to individuals and objects, has been used successfully in most parts of the world for about 25 years. Its purpose is the repeated identification of items for industrial and commercial purposes. Bar code consists of lines and spaces of varying width that contain data under different headings known as *symbologies*.

The great acceptance of these systems is due to their accuracy, precision and reliability in collecting printed data automatically and systematically and to their capacity for establishing unique communication and exchange links between mass-scale manufacturers and distributors of mass consumer products.

A. Start Up Procedures

- The national association for distributors or manufacturers interested in starting up a bar code system becomes a member of EAN (International Article Numbering Association), thus obtaining a 2 or 3 digit country identification code, known as FLAG. This number identifies the country of origin of each product. In the United States this association is the UNIFORM CODE COUNCIL (UCC).
- Manufacturing industries request from this national association in charge of code assignment, a set of numbers that will identify the company and will be the same for all items from the same company; next, separate numbers can be assigned to each item or item variation, with which a unique series of numbers for each item, known as CODE, is defined including:

COUNTRY + MANUFACTURER + ITEM + CONTROL

A 12-digit code known as UPC in the USA and Canada, or a 13-digit code called EAN elsewhere in the world, is thus formed. This code is composed of a number of vertical bars, or *symbol*, (for automatic scanning) plus a set of printed numbers, or *character set*, in human readable characters.

Manufacturers can use this same code to identify each item within their organization, in their internal systems of production, management, accounting, stock, sales and traffic.

— For exporting to countries outside the United States and Canada and for domestic use, the UPC code is used everywhere.
— Items imported into the United States or Canada will also need to carry a UPC code, because many United States scanners are still programmed this way.
— Outside the United States and Canada, all countries use the EAN bar code system, both for domestic and international needs.

• Manufacturers must design their packaging with the correct code location and order their packaging material from packaging manufacturers or printers, who should properly print and check the bar codes to comply with UPC/EAN specifications. For either small or large-scale production, coded labels can be attached to items.

All these processes require very strict standards and quality control that should never be omitted since out-of-specification printed codes prevent automatic marketing of items or, even worse, may cause costly return of packaging material or packed items. The consequences can be very expensive, particularly in export trade.

• Distributors, such as supermarkets or retail stores, adopt the UPC/EAN code of each item for identification in their internal network made up of various departments such as purchasing, stock, management, accounting, traffic and sales, using a central computerized system in different sectors of the organization linked to their checkout counters or *points of sale* (POS).

POS are equipped with a reading device known as a *scanner* on the countertop where a light beam, usually red laser, constantly scans in many directions at very high speed, exploring and analyzing data in three dimensions; this process is known as *scanning*.

Figure 2-1

B. The Scanner at the Point of Sale

When an item is scanned, the scanner recognizes the presence of an object within its visual field and activates a security window effect for as long as the object remains in that zone, thus preventing checking of the same item twice. This possibility depends on the type of scanner used. During the few seconds the employee needs to expose an item, the scanner can perform thousands of readings in different directions. The information is immediately collected and decoded by a computer until a specific interpretation is obtained, that is, the identification of a code number belonging to the bar code system for which it was programmed. Now the checkout computer searches for the item name and updated selling price in the memory system and prints them both on the customer receipt, where the purchase is indicated and totaled. A recently developed generation of "talking" scanners use a synthesized voice to tell item price, description, total amount and change due to the customer.

This information usually appears on an electronic screen placed at the checkout counter for visual control. *The code itself does not contain item prices*, thus avoiding individual price re-labeling in the event of a change in price. Selling price and item name are stored in the computer memory and can be modified when necessary by the retailer.

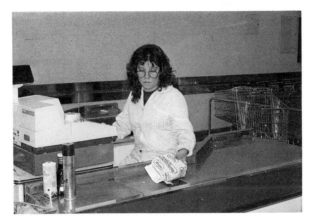

Figure 2-2

The computer also records sales into its accounting and stock volume system for an updated stock control and keeps statistical records per item.

This operation is performed in-line with each item being identified by the scanner at the point of sale.

Figure 2-3

C. Available Information

Information is processed and stored in computers on the basis of a binary digital language where everything is reduced to a series of ones and zeros. Both memory and central logical decision-making can easily be handled by a standard electronic computer compatible with most popular brands and models available everywhere. They can be interconnected or linked to branch offices and distributors for data centralization.

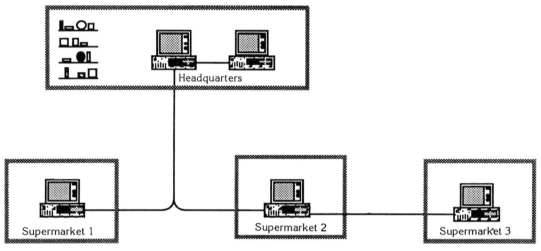

Figure 2-4

Now distributors can have a better knowledge of the dynamic parameters of their commercial circuits. Through this knowledge, distributors can improve performance and decision-making, as they will immediately and accurately be able to handle all data from retail outlets, whether located at the headquarters or not. Distributors will be able to know the stock period of every item and the daily habits of consumers and their shopping routines. In this way, distributors can plan times for special sales, which items to include and at what price.

Distributors will also know the items and brands consumers prefer, so they can restock supplies according to better selectivity criteria and spot the best moment to sell low-turnover goods.

Well-interpreted data can help identify less profitable items or items causing financial or economic losses. Equally important is the identification of branches and counters where these losses take place, and the times when they occur so that appropriate preventive measures may be taken.

Figure 2-5

If distributors share information with their suppliers, sales data, can also feed the production system by recommending changes after considering customer preferences for a given form of packaging, presentation, size, color or taste.

Today, the most advanced use of such data allows certain supermarkets to market complete generic item lines carryting their own brand in economy packaging, usually white, thus reducing the cost of advertising, packaging and low turnover, as shopping routines and item sale volumes are known in advance.

Figure 2-6

Whether they work on centralized supplying systems or not, multibranch supermarket chains can connect their computer systems on-line for data centralization and better strategic planning, since items that sell poorly in one branch may be consumer favorites in other areas. Centralized data allows top management to show branch managers the existing choices of items available in stock, and to propose promotions, items to add or delete, and so on.

Even though lack of information is inconceivable nowadays, thanks particularly to the broad scope of computerized systems available at reasonable prices, we must beware of the opposite extreme, i.e., excess of information. A great deal of information may prove worthless if the information cannot be properly interpreted and processed or if an adequate analysis of phenomena and routines cannot be made as they occur, in order to make immediate decisions. Also, regardless of data speed, gathering and analyzing data occupies time and space in the computer system, neither of which is free.

Adopting bar code systems in a business organization means learning a new language, which takes time, study and practice. Beginners should follow these general concepts:

- Only one person or working team should be responsible for the whole bar code project in the organization. This helps prevent a very common situation where many people might have only a small share of the required knowledge. Because of the systems complexity, the best thing is for very few people to know almost everything about bar codes.
- Acquire the maximum information in each area.
- Information on similar existing situations should be found.
- Contact should be made with every person and area within the company directly or indirectly associated with the system being adopted.
- Responsible suppliers should also be involved in every aspect concerning equipment, services and system, setup personnel, assistants, local UPC/EAN agency, item suppliers, printers, converters, packaging manufacturers and every area in any way related to bar codes.

- Bar code evolution.

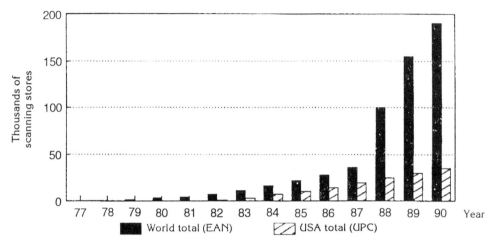

Figure 2-7 *Supermarkets and Retail Shops using scanners worldwide (UCC, EAN. 1990)*

D. System Benefits

- A single item ID from primary production sources to consumers, thus avoiding alterations and errors.
- Precise information on production times and cycles, inspection, storing, transportation and sales.
- General statistical data.
- Minimum data errors owing to auto check systems and control characters.
- Vertical oversize that allows a code to be scanned even if, as a result of damage, only a small part of its total height can be read.
- Faster, more efficient reception, sale and collection, particularly at supermarket checkout counters.
- No more price remarking or item-by-item relabeling, especially in countries with high-inflation.
- Prompt information on stock, sales and supplies.
- Human error elimination in marking, interpreting, invoice preparation and unknown loss.
- Adaptable to most packing and printing systems, and existing packaging materials and techniques.
- Easily adaptable and compatible with most available computer brands and systems.

Figure 2-8

E. Scanners Are Good Business, a Case Study

This is an analysis of the benefits derived from the installation of bar code scanners in cash register lines under standard conditions and was performed by a large domestic supermarket group whishing to reconfirm the convenience of scanners at their points of sale. We are indebted to Passeron and Barmat under the direction of Eng. F. Braun for this analysis.

Direct Benefits of Scanners at Point of Sale

In this case study, 80 percent of the items are coded at origin. This percentage corresponds to practically everything that can be coded at origin; percentages considered are slightly lower than those prevailing in the United States and Great Britain. Analyzed parameters include:

Store with 11 points of sale (registers)
Average purchase: 11 items per customer
Attending 44,000 customers per month
Items coded at origin: 80 percent
Scanner *first read rate*: 85 percent, second reading: 10 percent
Analysis: labor savings

- Time on the register line

The total daily operating time of a register consists of the cashier's productive and nonproductive times. Productive time is divided into *registering time*, corresponding to the period between pressing the first key and pressing the total key, and *collection time*, from the end of registering to the closing of the cash drawer. *Intermediate times*, corresponding to the pauses between successive customers in line, are included in collection time.

- Customer attention time

We call a register/item time, the result of dividing registering time by the number of items that entered the operation. In all the following examples, operating parameters are used which are appreciably lower than the theoretical operating limits established for a cashier's job.

Time savings presented corresponded exclusively to productive time. Therefore, given that the work is below the theoretical limit, there is an increase in rest time available for each cashier in addition to labor savings.

1. Direct Benefits, Productivity on Register Line

Case 1: Wage savings on a register line for a given number of customers.

Times:	Keyboard	Scanner	Scanner advantage
Registering time/item (sec.)	4.2	2.2	−47 percent
Collection time/customer (sec.)	29.0	29.0	
Customers/register per hour	20.0		
Theoretical limit	38.3		
Operating Results:			
Registration time/client (sec.)	46.2	24.2	
Collection time (sec.)	29.0	29.0	
Total time/customer	94.0	66.5	−29 percent
Working hours	4.2	3.3	

The registering time per keyboard (4.2 sec.) was taken from the register tapes of several stores over a period of five months (5 digit ID code). The scanner registering time (2.2 sec.) was taken from United States figures and verified locally in a pilot register line, with 85 percent scanner first read rate and an additional 10 percent on the second time.

- Economic Results, Savings

Time-saving/customer (sec.)	27.5
Hours saved/month	336.0
Hours saved/day	13.0
Hourly cost (US$/hr)	2.5
US$ saved/month	840
US$ saved/year	10,083
US$ saved/scanner year	1,008

Case 2: This is an extreme case of a store with limited available space and an unlimited number of customers.

Times:	Keyboard	Scanner	Scanner advantage
Register time/item (sec.)	4.2	2.2	
Collection time/customer (sec.)	29.0	29.0	
Store Parameters:			
Registers	11	11	
Items/customer	11	11	
Customers/month	44,000	62,195	+ 41 percent
Operating Results:			
Registration/customer (sec.)	46.2	24.2	
Collection time (sec.)	29.0	29.0	
Total time/customer	94.0	66.5	−29 percent
Working hours	4.2	4.2	

- Resulting profits, savings:

Time saved/customer (sec.)	27.5
Increased operations/month	18,195
US$/operation	9.9
Sales increase/month	180,130
Net income/month(US$)	9,006
Net income/year(US$)	108,078
US$ income/scanner year	9,825

2. Elimination of Labeling Procedures

Analysis of elimination of labeling work on products with a 5-digit numerical code: by taking measurements in several supermarkets, the *average* standard labeling time has been determined.

Labeling	1.41 sec.
Preparation	0.12 sec.
Other times	0.36 sec.
Standard time	1.89 sec.
Cost per worker hour	2.00 US$
Waste	10 percent
Articles/register/month (*)	48.400
type A items (bar-coded items)	80 percent
items to label per month/register	38,720
items to label/month	387,200
label consumption/month	424,600

(*) 10 registers, 11 items/customer, 49,000 customers/month

- Savings achieved, (in US$/year per scanner)

Labor	487	(35 percent)
Material	254	(65 percent)
Total	741	

3. Loss Reduction

- Quantifiable data:
  Involuntary keyboarding errors:

Keyboard input error	1 in 300
Scanner input error	1 in 3,000,000
Estimated amount of average error	−0.40 US$
Items/month/register	48,400
Items A (bar-coded items)	80 percent
US$ savings/scanner/year	581

Keyboard and scanner input errors have been taken from studies made in the United States which give coinciding values. The number of the average errors is an estimate based on the supposition that the majority of positive errors are detected and indicated by the customer, while the majority of negative errors go undetected. The percentage taken is 80 percent. Average item value is US$ 1.00. Also, as in the case of the supermarket studied, a 5-digit code with no control digit is used.

- Non-quantifiable losses
- Changes of labels (intentional or not).
  The existence of a bar code printed at the origin prevents the intentional changing of the label by the customer or personnel. Evaluation of this benefit is too uncertain to give a value estimate. We should mention that it could be insignificant.

- Theft in the register line (input of false codes).
  The scanner aids cashier supervision and makes it difficult to input substitute codes.
  As with the previous point, this is difficult to quantify.

- Cashier creativity stands out.

- Less merchandise manipulation in restocking (breakage).
  Even if this benefit exists it is difficult to evaluate. This is an important benefit for those looking for label excellence.

4. Other Direct Benefits

- Reduction store lines.
  In the productivity analysis we noted a reduction of one in the number of registers. Nevertheless, the theoretical value is 1.36. As it is impossible to subtract 0.36 checkouts, this difference is

transferred, given the same number of customers, in shorter checkout lines, increased exhibition space and better customer circulation.

- Improvement in working conditions
  According to studies conducted by the department of ergonomics of the University of Ohio, the installation of scanners in checkout counters considerably reduces the cashier's wrist movements. This takes on considerable importance in the United States since the average cost of "carpa tunnel syndrome" suits (wrist injury at work), was US$ 7,000 in 1989.

- One of the most unpleasant jobs, labeling, is eliminated.

F. Conclusion

Sooner or later, consumers will decide to shop where they can enjoy the benefits of scanners. Distributors, supermarkets and retailers will want these customers and the above-mentioned advantages. For this reason, they will market bar-coded items and will demand pre-coded items from their suppliers. This requires good design, printing and quality control of packaging bar codes. The main purpose of this book is to explain how this can be achieved.

G. Glossary of Bar Code Terms

ADD-ON: A second bar code symbol in addition to the main one.

AIM (AUTOMATIC IDENTIFICATION MANUFACTURERS): The Automatic Identification Manufacturers trade association represents manufacturers (vendors) of automatic identification equipment, systems, and supplies. This includes bar coding, radio frequency identification, magnetic strip, optical character recognition, voice recognition, and visual systems. The industry is a young, high-technology industry with technological developments originating primarily in the United States. AIM in the United States has served as an educational and marketing organization for the technology, providing educational programs that include speaker's bureaus, articles, publicity releases, and sponsorship of conferences and shows.

The AIM organization gives potential users of the technology confidence that the system is viable. The Auto ID technologies are basically peripheral to computerization; they need computer systems to function. Prior to AIM's international development effort, major manufacturers of Auto ID technologies were in the process of expanding their sale efforts to include dealers, distributors and representatives in other countries. (AIM 1988).

Figure 2-9   (*Courtesy of AIM*)

In 1983, AIM created the Technical Symbology Committee (TSC) to provide technical assistance for the development of generic bar code standards. TSC created five uniform symbology specifications (USS) to describe the most common symbologies for worldwide applications: USS 39, USS 12/5, USS Codabar, USS 128 and USS 93.

ANALYZER: An accurate, electronic lab instrument used in bar code quality control to prepare and check printing processes, to read, decode and analyze in detail different types of symbols used in packaging design, printing, printer quality control and reception of packed items. Generally suitable for color, reflectance and contrast analysis.

Figure 2-10   (*Courtesy of Stork Graphics*)

ASCII: Character code and group described in the American National Standard Code for Information Interchange. It is used for information exchange betwen communication and data processing systems.

ASPECT RATIO: Ratio of bar height to the symbol length.

BAR: Longer-than-wide line, the darker element of a printed bar code symbol, variable in width between one and several modules, able to absorb scanner light with minimum reflection.

Figure 2-11

BEARER BAR: Frame or outside border line formed by a thick, rectangular bar around the code, not interpreted by the scanner. It supports code bars to avoid quick, out-of-control, width enlargement in certain flexographic prints. Use is optional.

Figure 2-12

BI-DIRECTIONAL: Code capable of being read in both directions by the scanner, although later it will be electronically decoded in the right direction.

BINARY: Alphanumeric system using only two elements, "1" and "0".

BWR (BAR WIDTH REDUCTION): Reduction or enlargement of bars on a film master as a result of print tests using the *printability gauge* to *offset print gain*.

CHARACTER: Each number or letter symbolized as bars, spaces or algorithms, to be scanned. Each character is composed of as many 1's and 0's as the number of modules it contains.

Figure 2-13

CHECK DIGIT, CHECKSUM CHARACTER: A number included in the code, calculated by an algorithm based on remaining code numbers, used to ensure code is correctly composed.

CHARACTER SET: Encoded symbols printed in human readable characters, usually placed at the bottom of the code.

Figure 2-14

CODE LENGTH: Might be FIXED LENGTH: Symbol width is fixed, independent from coded information, like UPC and EAN codes; or VARIABLE LENGTH: Symbol width is proportional to the encoded information as in codes: 39, 93, 128 and Codabar.

COMPATIBILITY: The capacity a code has to be scanned and interpreted in a different system. The UPC system, for instance, is compatible with EAN, which can decode it, but EAN is not yet fully compatible with UPC, and might produce invalid readings if scanner is not specifically programmed.

EAN   UPC

Figure 2-15

CONTINUOUS CODE: A code where each character is next to the other, without inter character gaps. This means that all spaces are part of the code (the opposite of discrete code).

CONTRAST: The difference in color absorption between bars and spaces. The correct performance of a scanner is based on the recognition of contrast between element colors. Color and contrast conform to very exact specifications, and should never be determined without consulting a color guide.

CORNER MARKS: Marks or dots that limit the outside of a code and its elements, forming a rectangle inside which only code elements can be printed.

Figure 2-16

DENSITY: A code's density is the ratio between the number of encoded characters (modules) and the length they take once printed. Usually expressed in "characters/inch (cpi)" or "characters/cm", "modules/cm" or "X/inch". Density depends directly on module, wide to narrow ratio, code type and printing system; and is classified into three categories according to module width, as follows:

High density: Module below 0.01″   (.254 mm)

Medium density: Module between 0.01″ and 0.02″ (.254 - .508 mm)

Low density: Module above 0.02″   (.508 mm)

Figure 2-17   *High density symbol*

Figure 2-18   *Low density symbol*

DEPTH OF FIELD: The difference between maximum and minimum readable distance from the scanner to the symbol. Applies only to scanners that do not require physical contact with the symbol's printed surface.

DIGIT: Each of the numbers or alphabetic characters, symbolized in bars and spaces, different in value and representation a form.

DISCRETE CODE: A code where each character is independent and separated from the other by an inter character gap that is not a part of the code (the opposite of continuous code).

EAN (INTERNATIONAL ARTICLE NUMBERING ASSOCIATION): Representatives from 12 European countries gathered in 1973 to discuss a unified encoding method and signed the EAN Agreement

Memorandum on February 3, 1977. This is officially considered as the date of EAN's origin. The 12 founding countries did not anticipate the outstanding growth EAN would experience in the coming years, and they named it "European Article Numbering" (EAN). The name was then changed to "International Article Numbering Association" in 1981. However, the original initials were maintained to identify the symbology and numbering system currently applied all over the world for encoding commercial items, designed to be compatible with the UPC symbology used in the United States.

Nowadays, EAN has 55 member countries on all continents, representing more than 200,000 scanning stores and 280,000 companies affiliated to local EAN organizations in most countries. EAN headquarters are in Brussels, Belgium, where the European Economic Community is located. (EAN 1988,1993)

Figure 2-19

FILM MASTER: The original film on which a code is recorded for the first time by computerized photosetting equipment. The code thus produced must be perfect to prevent future error accumulation in photographic processes when reduction and enlarging processes are involved prior to final printing.

Film master always contains all bars, spaces and character sets and generally includes identification references such as code type, customer's name, item name, MF, BWR, production date and other useful information.

Figure 2-20   (*Courtesy of Stork Graphics*)

FIRST-READ RATE: The rate of correct readings the scanner will produce on its first scan.

FLAG: Code reference assigned to a country or type of item to identify items internationally. Used basically in the EAN system, it consists of two or three digits. The UPC equivalent is the *number system character* that indicates item type and has only one digit.

GUARD BARS PATTERN: Auxiliary characters formed by bars and spaces that usually warn scanner about code start and stop. Guard bars are also used within the code to separate zones.

INTER CHARACTER GAP: The space separating one character from another in discrete codes that are not part of the code. Inter character gaps do not exist in continuous codes.

INVALID SCAN: Occurs when scanner reads a code but does not recognize it as such.

LASER: A light beam traveling coherently in phase without spreading out, resulting in high-energy concentration. Artificially created from high molecular stimulation in a gas medium (e.g., helium-neon laser), a crystal (e.g., ruby or artificial crystal) or a solid-state semiconductor called light-emitting diode (LED). Beam frequency determines color and visibility. Scanners may use visible red or invisible infrared laser, since their wavelength (or frequency) is such that it will be absorbed by the bars and reflected by the spaces of the bar code. Although it is possible to obtain high energy concentrations in lasers, their use in scanners should not concern us as it is harmless, since it is a low intensity beam free of harmful radiation, as long as manufacturer instructions are followed.

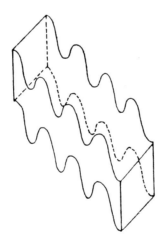

Figure 2-21 *In-phase coherent light beam called Laser*

LED (LIGHT-EMITTING DIODE): A transistor-like solid state device capable of transforming small amounts of electricity into light energy. A LED is a long-life transducer, usually made of gallium arsenide. Scanner LEDs usually emit in different wavelengths and propagation forms. There are noncoherent LEDs, usually red, and coherent (laser), red or infrared. LEDs are widely used in all types of scanners.

MAGNIFICATION FACTOR (MF): Adopting the standard dimensions of a symbol the magnification factor or standard size is MF = 1. It is possible to increase its relative size to MF = 2.0 maximum or reduce it to MF = 0.8 minimum (i.e., up to 200 percent increase and no less than 80 percent reduction, taking 100 percent as normal size).

For EAN and UPC codes, these MF limits, 0.8 and 2.0, should not be exceeded. Printing an MF > 1 is specially recommended whenever possible, since printing tolerance is reduced by 66 percent when a code is compressed to MF = 0.8, which makes it difficult to keep within printing specifications.

The decision for a low MF should not be arbitrary and must depend on the results of a printing test performed with the *printability gauge* and other variables according to code's design and packaging.

Figure 2-22

MINIMUM REFLECTANCE DIFFERENCE (MRD): The smallest rate (for a specific range of wavelengths) between the amount of light reflected by a surface (the code) and the amount reflected by a barium oxide or magnesium oxide standard pattern.

MODULE ("X" DIMENSION): The narrowest element, either a bar or a space, in a bar code. All code elements including quiet zones and *guard bars* have a module multiple as their width. Module size (width) directly defines density and is the "X" dimension, or nominal dimension, of a code.

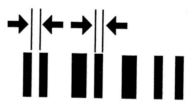

Figure 2-23 *Light or Dark Module*

NORMAL SIZE: The normal or standard bar code size corresponds to a magnification factor MF = 1 (often indicated as 100 percent). It includes all the area inside the *corner marks*.

PORTABLE SCANNER: Hand-operated portable unit that can be taken to the item to be scanned. (See chapter 10)

Figure 2-24

PRINT GAIN: Increase (or decrease) in printed bar size compared to original film master.

PRINTING RANGE: Set of letters (A - K or A' - K') resulting from printing trials using Printability Gauge.

PRINTABILITY GAUGE: Consists of parallel lines arranged in 11 sets identified with letters A - K longitudinally and A' - K' across, on a special photographic film.
Lines are closer in each set, and the gauge is used by the printer to determine printed symbol size (MF), BWR and corresponding *print gain*.

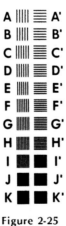

Figure 2-25

QUIET ZONES: Reserved zones or margins with no bars, composed only of spaces and placed to the left before start character and to the right after stop character. Quiet zones are usually about 10 modules wide each, according to code. Some of these quiet zones, together with the bar and space pattern defined, allow scanner to recognize a code as such. If these margins are reduced, scanner will fail to interpret the rest as a code, producing an invalid scan.

Figure 2-26

REFLECTANCE: Ratio between incoming and reflected light flow.

SCANNER: Light-emitting and —receiving transducer able to turn printed data into digital pulses able to feed a computer. The word *scanner* is used to describe the instrument and/or person performing the actions of reading, exploring and analyzing a printed bar code.

SPACE: Longer-than-wide line, the whitest element of a printed bar code symbol, variable in width between one and several modules, able to reflect scanner light without absorbing it. Usually forms the background on which bars are printed.

STANDARD DIMENSION: Code length and surface when magnification factor MF = 1 (or 100 percent).

START CHARACTER: Indicates beginning of code to scanner. Depending on the code, it can be either a number, letter or symbol.

STOP CHARACTER: Indicates end of code scanner. Depending on the code, it can be either a number, letter or symbol.

Figure 2-27

SUBSTRATE: Packaging or label surface on which bar codes are printed.

SYMBOL: Specific bar and space arrangement in the code. Each symbol stores data in its own form, different from the rest, in a language that will only be interpreted by a scanner.

Figure 2-28

TOLERANCE: Tolerance or dimension errors must always be kept within UPC/EAN specifications and as low as possible because in subsequent processes they may accumulate to a degree that might cause an invalid scan. When a code is enlarged or reduced, so is its tolerance. The rule is "the greater the magnification factor the higher the tolerance, and the greater the reduction, the lower the tolerance".

UNIFORM CODE COUNCIL INC. (UCC): Organization responsible for administration of UPC and other symbols for article numbering in the United States.

UPC (Universal Product Code) Standard bar code symbol in the United States and Canada for retail packages. UPC is administered by UCC in the United States.

Figure 2-29

VOIDS: White or clear spots in the symbol caused by poor printing quality; usually ink is missing due to variations in viscosity or lack of pressure on printing cylinders or plates are due to be replaced.

Figure 2-30

WIDE-TO-NARROW RATIO: Relationship between width of widest and narrowest elements in the code.

3

Bar Codes
in the United States
and Canada

UPC, or Universal Product Code, the most popular bar code, was created and adopted by United States industry in 1973 to be used at point of sale, mainly in supermarkets and retail stores. There are two available options for this symbol: UPC-A and UPC-E.

A. UPC-A Code

1. UPC-A Characteristics

Characters: A total of 12, numbers only, as follows:
Character 12:   Number System Character
Character 11, 10, 9, 8, 7:   Manufacturer's ID
Character 6, 5, 4, 3, 2:   Item ID
Character 1:   Check Digit

Figure 3-1   *UPC-A code, MF =1*

Characters are single-digit numbers (0 1 2 3 4 5 6 7 8 9), 12 of which will form the UPC-A code (as an example only, code 012345678905 is used in this chapter and diagrams).

The twelve characters forming the code will be represented and printed as bars and spaces (symbol) for scanner reading purposes. Characters 1 and 12 are printed with longer bars than the others.

Each character is represented by two bars and two spaces alternately arranged, i.e., 4 elements per character; element width and location make the difference between one character and the other.

Character width is fixed, measuring 7 modules (a module is the thinnest element). Therefore, all four elements of a character will have a total width of 7 modules, so each bar and/or space will be 1 module wide minimum and 4 module wide maximum, thus forming a complex structure code.

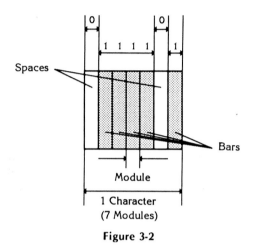

Figure 3-2

The above criteria only apply to the 12 characters encoded in UPC-A system and do not apply to guard bars or quiet zones.

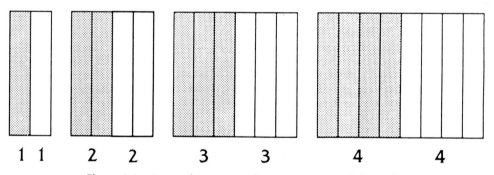

Figure 3-3 *Bars and Spaces can be 1, 2, 3 or 4 modules wide*

For a better understanding of this text, character location or positioning in the code is defined as follows: Position 1 the first character to the right of code; position 12, the last character to the left of code. This method of locating characters is only for study purposes and is not part of the codification.

This apparent complexity has the purpose of allowing the computer decoder to identify the code, finding the correct direction in bi-directional reading, activating electronic checking mechanisms, avoiding errors and preventing wrong scans. However, these complications should not frighten users, since the film masters needed for symbol printing are computer-generated. Handmade design and drawing are useless and unnecessary. Film master computerized photosetting only requires the complete number assigned, size (MF), code type and a few photographic details each printer is able to obtain.

- Guard bar's pattern:
    Left, fixed width: 3 modules, BAR-SPACE-BAR, encoded 101
    Right, fixed width: 3 modules, BAR-SPACE-BAR, encoded 101
    Center, fixed width: 5 modules, SPACE-BAR-SPACE-BAR-SPACE, encoded 01010
    Guard bar's standard height: 0.965" (24.5 mm), slightly higher than most bars
- Quiet zones:
    Left: 9 modules wide minimum, encoded 000000000
    Right: 9 modules wide minimum, encoded 000000000
    Lower: 1 module, between bars and character set
- Encoding:
    Continuous, bi-directional
- Character set:
    Location: At the bottom of code, all characters in position 2 through 11
    At left quiet zone: position 12
    At right quiet zone: position 1 (optional)
    Character font type: OCR-B
- Structure: Complex
- Printed module, standard (MF = 1):
    Width: 0.013" ± 0.004" (0.33 mm ± 0.101 mm)
- Fixed length:
    113 modules between corner marks
    95 modules between right and left guard bar patterns
- Standard density: Medium
- Standard size (MF = 1)
    1.469" × 1.020" (37.3 × 25.9 mm) between corner marks,
    including quiet zones (right, left and lower)
- Symbol height (bar or space):
    0.900" (22.8 mm) except for characters in positions
    1 & 12, which are longer, the same as guard bars
- Magnification factors:
    MF = 0.8 to 2.0 (80 percent to 200 percent)
- Compatibility: Can be understood by EAN systems, where one character = 0 will be interpreted in position 13 (see Table 4-2).

- Code administration:

  Uniform Code Council, Inc. (UCC) in the United States. Local EAN agency elsewhere worldwide
- Number system character: (UCC 1990)

0	Regular UPC codes (version A and E).
2	Random weight items, such as meat and produce, symbol marked at store level (Version A)
3	National Drug Code (NDC) and National Health-Related Items (HRI) Code in current 10-digit code length (version A). Note that symbol is not affected by the various internal structures possible with the NDC or HRI codes.
4	For use without code format restriction and with check digit protection for in-store marking of non-food items (version A).
5	For use on coupons (version A).
6,7	Regular UPC codes (version A).
1,8,9	Reserved for uses unidentified at this time.

2. UPC-A General Diagram

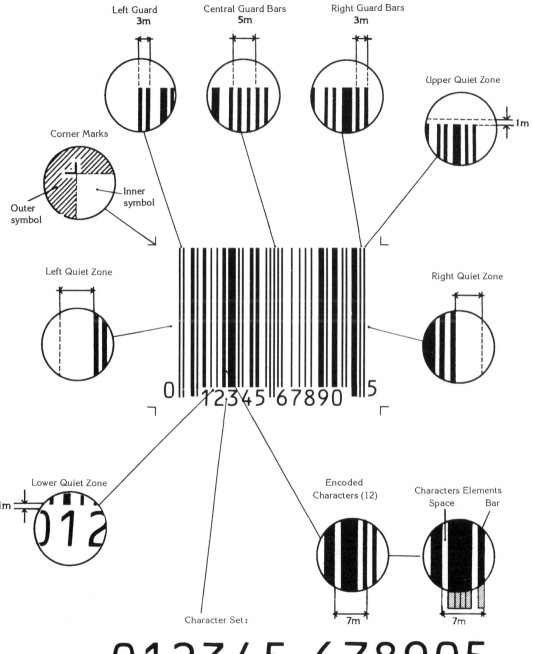

Left Guard
3m

Central Guard Bars
5m

Right Guard Bars
3m

Upper Quiet Zone

1m

Corner Marks

Outer
symbol

Inner
symbol

Left Quiet Zone

Right Quiet Zone

Lower Quiet Zone

1m

Character Set:

Encoded
Characters (12)

Characters Elements
Space   Bar

7m

7m

0   1 2 3 4 5   6 7 8 9 0   5

012345 678905

Figure 3-4

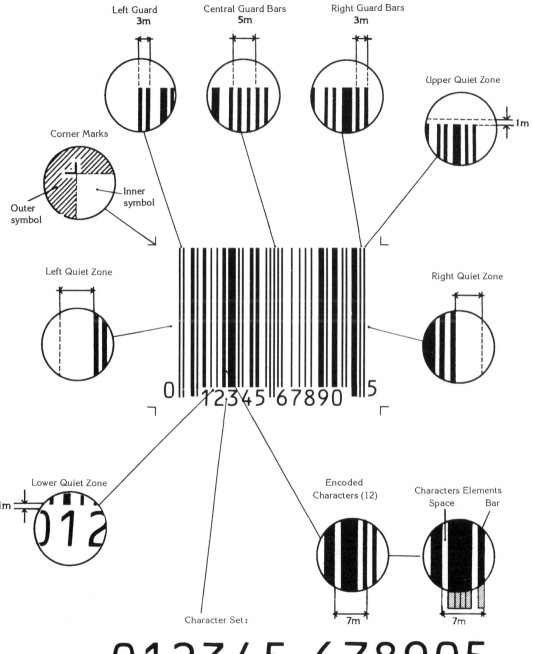
33

3. UPC-A/E Magnification Factors and Dimensions

BRITISH

MAGNIFICATION	Inches				Inches/1000		Inches/1000				
	Left Edge of First Bar to Right Edge of Last Bar		Total Width Incl. Margins		7 Mods.	1 Mod.	Nominal Margin — Left (From Left Edge of 1st Bar)	Nominal Margin — Right (From Right Edge of Last Bar)		Min. Guard Bar Ht.	Bar or Space Width Toler. ±
	Version A	E	Version A	E	Vers. A/E	Vers. A/E	Vers. A/E	A	E	Vers. A/E	Vers. A or E
0.8	.988	.530	1.175	0.697	72.8	10.4	93.6	93.6	72.8	768	1.4
0.9	1.112	.597	1.322	0.784	81.9	11.7	105.3	105.3	81.9	864	2.7
1.0	1.235	.663	1.469	0.871	91.0	13	117.0	117.0	91.0	960	4.0
1.1	1.359	.729	1.616	0.958	100.1	14.3	128.7	128.7	100.1	1056	4.6
1.2	1.482	.796	1.763	1.045	109.2	15.6	140.4	140.4	109.2	1152	5.2
1.3	1.606	.862	1.910	1.132	118.3	16.7	152.1	152.1	118.3	1248	5.8
1.4	1.729	.928	2.057	1.219	127.4	18.2	163.8	163.8	127.4	1344	6.4
1.5	1.853	.995	2.204	1.307	136.5	19.5	175.5	175.5	136.5	1440	7.0
1.6	1.976	1.061	2.350	1.394	145.6	20.8	187.2	187.2	145.6	1536	7.6
1.7	2.100	1.127	2.497	1.481	154.7	22.1	198.9	198.9	154.7	1632	8.2
1.8	2.223	1.193	2.644	1.568	163.8	23.4	210.6	210.6	163.8	1728	8.8
1.9	2.347	1.260	2.791	1.655	172.9	24.7	222.3	222.3	172.9	1824	9.4
2.0	2.470	1.326	2.938	1.742	182.0	26	234.0	234.0	182.0	1920	10.1

METRIC

MAGNIFICATION	MM						MM				
0.8	25.8	13.46	29.83	17.69	1.848	.264	2.38	2.38	1.85	19.50	.036
0.9	28.21	15.15	33.56	19.90	2.079	.297	2.67	2.67	2.08	21.93	.069
1.0	31.35	16.83	37.29	22.11	2.310	.330	2.97	2.97	2.31	24.37	.100
1.1	34.48	18.51	41.02	24.32	2.541	.363	3.27	3.27	2.54	26.81	.116
1.2	37.62	20.20	44.75	26.53	2.772	.396	3.56	3.56	2.77	29.24	.131
1.3	40.75	21.48	48.48	28.74	3.003	.429	3.86	3.86	3.00	31.68	.147
1.4	43.89	23.56	52.21	30.95	3.234	.462	4.16	4.16	3.23	34.12	.162
1.5	47.02	25.25	55.93	33.17	3.465	.495	4.46	4.46	3.47	36.55	.178
1.6	50.16	26.93	59.66	35.38	3.696	.528	4.75	4.75	3.70	38.99	.193
1.7	53.29	28.61	63.39	37.59	3.927	.561	5.05	5.05	3.93	41.35	.208
1.8	56.43	30.29	67.12	39.80	4.158	.594	5.35	5.35	4.16	43.86	.224
1.9	59.56	31.98	70.85	42.01	4.389	.627	5.64	5.64	4.39	46.30	.240
2.0	62.70	33.66	74.58	44.22	4.620	.660	5.94	5.94	4.62	48.74	.255

Table 3-1 *Dimensions are Inches/1000 and MM. (UPC SSM 1990)*

4. UPC-A Format and Dimensions

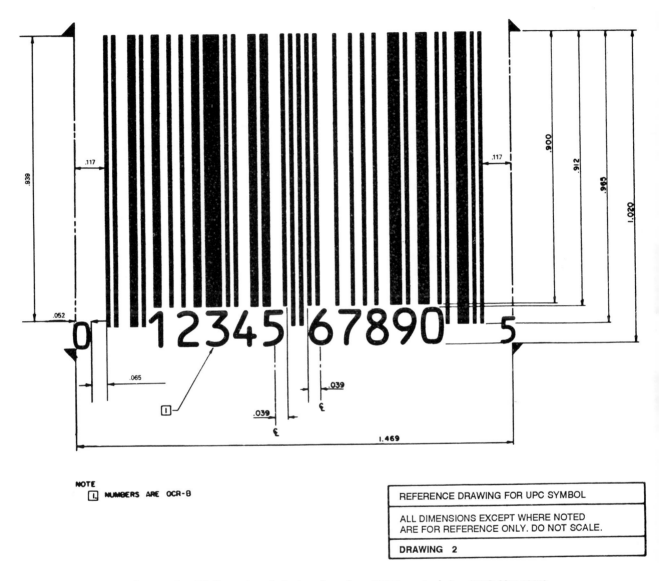

NOTE

☐ NUMBERS ARE OCR-B

REFERENCE DRAWING FOR UPC SYMBOL

ALL DIMENSIONS EXCEPT WHERE NOTED
ARE FOR REFERENCE ONLY. DO NOT SCALE.

DRAWING 2

Figure 3-5 *All dimensions in inches, based on 100% nominal size. (UPC SSM 1990)*

5. How to Obtain a UPC Code

United States manufacturers as well as foreign companies exporting their items to the United States and Canada will need a properly printed UPC code. To apply for it, they must become members of UCC, who will assign them a five-digit manufacturer ID number, and it will be up to them to assign another five-digit item ID number to each item. UCC membership fees may vary from $ 300 to $ 10,000 according to corporate activity and annual sales volume.

For more information, please contact your local EAN agency, or, if in the United States:

> Uniform Code Council, Inc.
> P.O.Box 1244
> Dayton, Ohio 45401-1244
> Fax: (513) 435-4749

6. A & C Character Encoding

There are two ways to encode characters located in positions 1 to 12, these are A and C, depending on bar width of each character, whether there is an odd number of modules (3 or 5), or an even number (2 or 4); and depending on whether the first and last module (of the seven forming each character) are a space and a bar, or a bar and a space, respectively:

Character A

> Has two bars composed of 3 or 5 modules (odd number)
> First left module: a space
> Last right module: a bar
> Located to the left of the center guard bar on positions 7 through 12

Character C

> Has two bars composed of 2 or 4 modules (even number)
> First left module: a bar
> Last right module: a space
> Located to the right of the center guard bar on positions 1 through 6

"A" Characters are located to the left of the center guard bar, and "C" characters are located to the right.

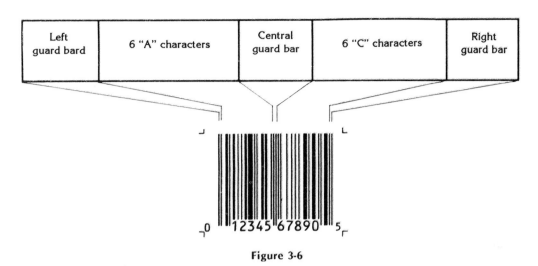

Figure 3-6

7. Encoding UPC-A/E: A & C Characters

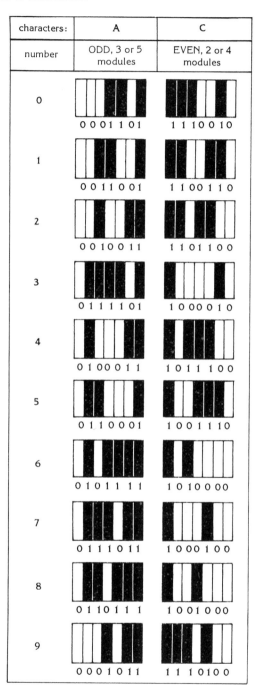

characters:	A	C
number	ODD, 3 or 5 modules	EVEN, 2 or 4 modules
0	0 0 0 1 1 0 1	1 1 1 0 0 1 0
1	0 0 1 1 0 0 1	1 1 0 0 1 1 0
2	0 0 1 0 0 1 1	1 1 0 1 1 0 0
3	0 1 1 1 1 0 1	1 0 0 0 0 1 0
4	0 1 0 0 0 1 1	1 0 1 1 1 0 0
5	0 1 1 0 0 0 1	1 0 0 1 1 1 0
6	0 1 0 1 1 1 1	1 0 1 0 0 0 0
7	0 1 1 1 0 1 1	1 0 0 0 1 0 0
8	0 1 1 0 1 1 1	1 0 0 1 0 0 0
9	0 0 0 1 0 1 1	1 1 1 0 1 0 0

Figure 3-7 *Each black module = 1, each white module = 0*

38

8. UPC-A Nominal Dimensions and Encoding Chart

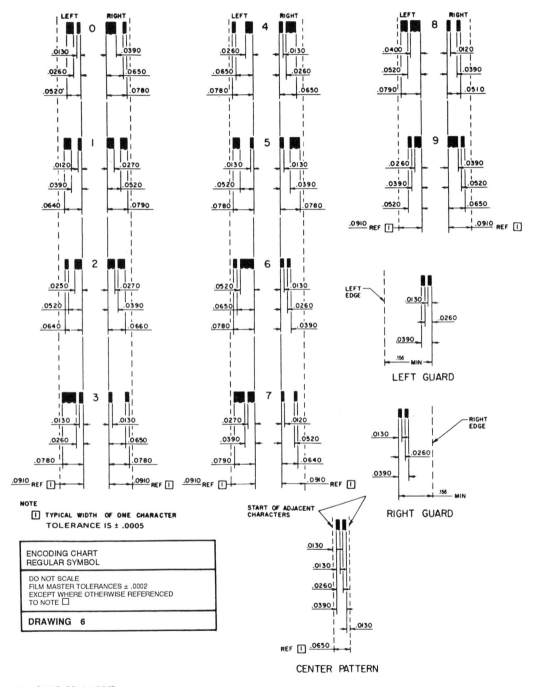

Figure 3-8 (*UPC SSM 1990*)

9. UPC Check Digit Calculation

Unlike the others, the check digit, located in position 1, is not assigned to an item but is the result of a calculation involving numbers located in positions 2 through 12.

The check digit is designed to prevent reading errors and wrong scanning usually caused by defective symbol design or printing that allow unassigned numbers to be read. The check digit also features a self-checking operation to detect self-calculation errors. Check digit calculation method is the same for UPC-A and EAN-13 codes, as follows:

(a) Multiply each character in an odd position X 1.
Multiply each character in an even position X 3.
(b) Add the 11 numbers obtained, getting a value known as "products sum" (PS).
(c) Divide the products sum by 10 (a constant), obtaining a quotient (C) and a balance (B).
PS/10 = C and B.
(d) Deduct the balance from 10, obtaining the check digit (CD) to be placed on position 1
10 − B = CD

- Check digit (CD) calculation using UPC-A code, number 012345678905:

Position #		12	11	10	9	8	7	6	5	4	3	2	1
	value:	0	1	2	3	4	5	6	7	8	9	0	?
	X												
(a)		3	1	3	1	3	1	3	1	3	1	3	
	=												
(b)	PS	0 +	1 +	6 +	3 +	12 +	5 +	18 +	7 +	24 +	9 +	0	

(b) PS = 85

(c) $\dfrac{PS}{10} = C \text{ and } B$ so, $\dfrac{85}{10} = 8 \text{ and } B = 5$

(d) 10 - B = CD so, 10 - 5 = $\boxed{5}$

(d) so Check Digit CD = 5 and we can place it in position # 1

Table 3-2

- Self-cheking

Self-cheking of CD character just calculated is performed using exactly the same procedure as above, with the following changes:

(a) The value in position 1 now exists and is the one calculated above, equal to 5; here the multiplying factor is always 1.

(c) If quotient results in zero balance (B = 0), character value calculated above is confirmed and self-checked in position 1.

- Check digit (CD) verification example using UPC-A code number 012345678905

Position #		12	11	10	9	8	7	6	5	4	3	2	1
(a)	value: × = so,	0 3	1 1	2 3	3 1	4 3	5 1	6 3	7 1	8 3	9 1	0 3	5 1
(b)	PS	0 +	1 +	6 +	3 +	12 +	5	+ 18 +	7	+ 24 +	9	+ 0 +	5

(b)   PS   = 90

(c) $\dfrac{PS}{10}$ = C and B   so, $\dfrac{90}{10}$ = 9 and B = $\boxed{0}$ check digit has been verified

Table 3-3

B. UPC-E Code (Reduced Code)

This version is known as "zero suppressed" as it eliminates at least 4 zeros in the UPC code. As a very simple example, UPC-A code 12300-00045 would turn out to be UPC-E 12345 thus reducing the symbol-required printing area. Using this version is not always possible depending on the manufacturer's number and item number. It is widely-belieed that UPC-E version consists of just taking zeros away from UPC-A code, but this is not so, as there are four ways of suppressing zeros, depending on the numbers assigned to manufacturer and item. In order to zero suppress a UPC-A code, the system character must necessarily be 0.

Proper application is subject to strict regulations that determine the number of items eligible for a UPC-E code in every case, as follows:

a— Manufacturer's number ends in 000, 100 or 200;
   1000 Items may be encoded under UPC-E (000 to 999).
b— Manufacturer's number ends in 300, 400, 500, 600, 700, 800 or 900
   100 Items may be encoded (00 to 99).
c— Manufacturer's number ends in 10, 20, 30, 40, 50, 60, 70, 80 or 90
   10 Item numbers can be assigned (0 to 9).
d— Manufacturer's number does not end in zero;
   only 5 items may use the reduced version (5 to 9).

In any case, the number assigned by the manufacturer to each item should start with some "zeros", if possible, in order to make code reduction easy when needed. This UPC-E reduced version of UPC-A code is based on a setup concept different from the one that allows assigning an EAN-8 code instead of a EAN-13, and should not be confused with it.

Figure 3-9   *UPC-E code, MF = 1*

1. UPC-E Characteristics

- Characters: 8 (although 12 will be read) numbers only.
   Four of the encoded characters follow "A" type sequence, and four follow "B" type sequence (see Figure 4-7).
   Manufacturer's and item characters will be encoded by zero suppression from the original UPC-A symbol as explained above. Each character consists of 2 bars and 2 spaces of 1 to 4 modules width each, the same as in UPC-A.
- Guard bars pattern:
   Left, fixed width: 3 modules, BAR-SPACE-BAR, encoded 101
   Right, fixed width: 5 modules, BAR-SPACE-BAR-SPACE-BAR, encoded 10101
   This encoding will indicate end of scanner reading, no center guard bar needed.
   Guard bar's standard height: 0.965" (24.5 mm), slightly higher than most bars.

- Quiet zones:
    Left: 9 modules wide minimum, encoded 000000000
    Right: 7 modules wide minimum, encoded 0000000
    Lower: 1 module, between bars and character set
- Encoding: continuous, bi-directional
- Character set:
    Location: At the bottom of code all characters in position 2 through 7
    At left quiet zone: position 8
    At right quiet zone: position 1 (optional)
    Character font type: OCR-B
- Structure: complex
- Printed module, standard (MF = 1):
    Width: 0.013″ ± 0.004″ (0.33 mm ± 0.101 mm)
- Fixed length:
    80 modules between corner marks
    64 modules between right and left guard bar patterns
- Standard density: medium
- Standard size (MF = 1)
    0.871″ × 1.020″ (22.1 × 25.9 mm) between corner marks,
    including quiet zones (right, left and lower)
- Bar height: 0.900″ (22.8 mm) except for guard bars
- Magnification factors: MF = 0.8 to 2.0 (80 percent to 200 percent)
- Compatibility: Can be decoded as a UPC-A code and understood by EAN systems
- Application: Whenever printed symbol size must be reduced to a minimum because of lack of space, provided original Symbol UPC-A allows for zero suppression
- Code administration: Uniform Code Council, Inc. (UCC) in the United States, Local EAN agency elsewhere worldwide

2. UPC-E Nominal Dimensions and Encoding Chart

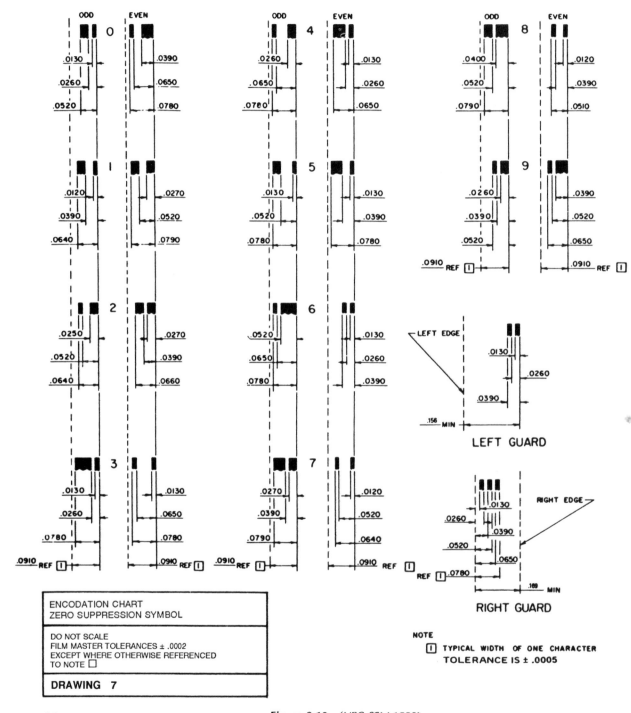

Figure 3-10   *(UPC SSM 1990)*

C. UPC Code Truncation

"Truncation" means cutting bar height up to a fraction of an inch in the upper side of the symbol. UPC symbol specifications *do not permit truncation*, this choice is only used as a last resort when available printable area is too small and code structure does not allow the use of UPC-E reduced version. A truncated symbol should never be simultaneously with low MF, as this would further reduce successful scanning chances.

Truncation strongly reduces or nullifies omnidirectional scanning capabilities and should be avoided.

4

EAN International
Bar Codes
(Outside United States
and Canada)

EAN is the bar code symbology for commercial use at point of sale (POS) all over the world, except in the United States and Canada, where UPC symbology is used. Both symbols are almost identical, but EAN permits encoding one extra character (position 13) outside the symbol, for country ID purposes.

This symbol is composed of a series of parallel dark bars and light spaces, varying in width. The system has a fixed number of bars (30) and spaces (29) exactly as the UPC code does.

In theory, the system allows encoding up to 1,000 countries or member organizations to include 10,000 different manufacturers with 100,000 items or item presentations each, which represents a huge number of possible combinations. There are two available options for this symbol: EAN-13 and EAN-8.

As the system is common to all countries and products within and outside the European Common Market, it logically bears an indicative ID reference known as a "flag" unique to every country or national EAN encoding organization responsible for local code assignment (See EAN flags chart, Table 4-10).

Figure 4-1   *EAN-13 code, MF = 1*

A. EAN-13 Code

1. EAN-13 General Diagram Figure 4-2

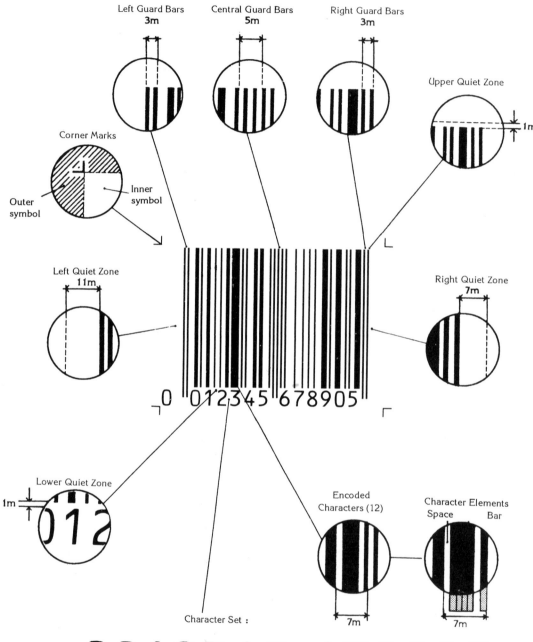

Left Guard Bars
3m

Central Guard Bars
5m

Right Guard Bars
3m

Upper Quiet Zone

1m

Corner Marks

Outer symbol

Inner symbol

Left Quiet Zone
11m

Right Quiet Zone
7m

Lower Quiet Zone

1m

Character Set :

Encoded Characters (12)

7m

Character Elements
Space   Bar

7m

0012345 678905

2. EAN-13 Characteristics

Characters: A total of 13, numbers only, as follows:

Character. 13:   Country ID (Flag) Not symbolized with bars and spaces
Character. 12:   Country ID (Flag)
Character. 11:   Country ID (Flag) or manufacturer ID
Character. 10, 9, 8, 7:   Manufacturer's ID
Character. 6, 5, 4, 3, 2:   Item ID
Character. 1:   Check Digit

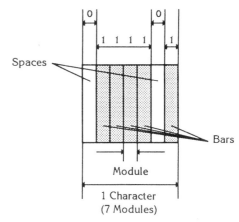

Figure 4-3

The characters are single-digit numbers (0 1 2 3 4 5 6 7 8 9), 13 of which will form the EAN-13 code (as an example only, code 0012345678905 is used in this chapter and diagrams). Twelve characters will form the symbol represented and printed as bars and spaces, but the character on position 13 will not.

Each character is represented by 2 bars and 2 spaces alternately arranged, i.e., 4 elements per character; element width and location make the difference between one character and another.

Character width is fixed, measuring 7 modules (a module being the thinnest element). Therefore, all four elements of a character will have a total width of 7 modules, so each bar and/or space will be 1 module wide minimum and 4 modules wide maximum, thus forming a complex structure code.

The above criteria apply only to the 12 characters encoded in EAN-13 system and do not apply to guard bars, quiet zones, or to the character in position 13.

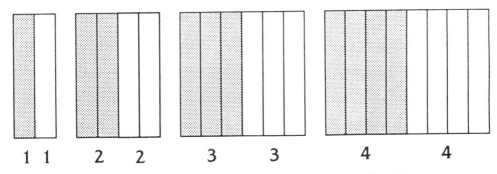

Figure 4-4 *Bars and spaces can be 1, 2, 3 or 4 modules wide*

For a better understanding of this text, character location, or positioning in the code, is defined as follows: Position 1 the first character to the right of code and position 13 the last character to the left of code. This way of locating characters is only for study purposes and is not part of the codification.

This apparent complexity has the purpose of allowing a computer decoder to identify the code, finding the correct direction in bi-directional reading, activating electronic checking mechanisms, avoiding errors and preventing wrong scans. However, these complications should not frighten users, as the film masters needed for symbol printing are computer-generated. Handmade design and drawing are useless and unnecessary. Film master computerized photosetting only requires the complete number assigned, size (MF), code type and a few photographic details each printer is able to obtain.

- Guard bars pattern:
    Left, fixed width: 3 modules, BAR-SPACE-BAR, encoded 101
    Right, fixed width: 3 modules, BAR-SPACE-BAR, encoded 101
    Center, fixed width: 5 modules, SPACE-BAR-SPACE-BAR-SPACE, encoded 01010
    Guard bars standard height: 24.50 mm (0.965''), slightly higher than most bars
- Quiet zones:
    Left: 11 modules wide minimum, encoded 00000000000
    Right: 7 modules wide minimum, encoded 0000000
    Lower: 1 module, between bars and character set
- Encoding:
    Continuous, bi-directional
- Character set:
    Location: At the bottom of code, all characters in position 1 through 12
    At left quiet zone: position 13
    Characters font type: OCR-B
- Structure: complex
- Printed module, standard, (MF = 1):
    Width: 0.330 mm ± 0.101 mm (0.013'' ± 0.004'')
- Fixed length:
    113 modules between corner marks
    95 modules between right and left guard bar patterns
- Standard density: medium
- Standard size (MF = 1)
    37.29 × 25.87 mm (1.47'' × 1.02'') between corner marks,
    including quiet zones (right, left and lower)
- Symbol height (bar or space):
    22.8 mm (0.90'') except for guard bars
- Magnification factors:
    MF = 0.8 to 2.0 (80 percent to 200 percent)
- Compatibility: Can be understood by some UPC systems if programmed
- Code administration: EAN local organization
- Flag: The flag or country ID number is 2 or 3 digits wide, position 13,12 or 13,12,11.
    See EAN flags chart, Table 4-10

3. EAN-13 Magnification Factors and Dimensions

EAN-13 Magnification Factors		Module width (M)	Between first and last bar			Between	Corner	Marks
			width	length	surface	width	length	surface
(MF)	(%)	mm	mm	mm	cm²	mm	mm	cm²
0.80	80	.264	25.08	19.60	4.916	29.83	21.01	6.267
0.85	85	.281	26.65	20.82	5.549	31.70	22.32	7.075
0.90	90	.297	28.22	22.05	6.221	33.56	23.63	7.932
0.95	95	.314	29.78	23.27	6.932	35.43	24.95	8.838
1.00	100	.330	31.35	24.50	7.681	37.29	26.26	9.792
1.10	110	.363	34.48	26.95	9.294	41.02	28.89	11.849
1.20	120	.396	27.62	29.40	11.060	44.75	31.51	14.101
1.30	130	.429	40.76	31.85	12.980	48.48	34.14	16.549
1.40	140	.462	43.89	34.30	15.054	52.21	36.76	19.193
1.50	150	.495	47.02	36.75	17.282	55.94	39.39	22.033
1.60	160	.528	50.16	39.20	19.663	59.66	42.02	25.068
1.70	170	.561	53.30	41.65	22.197	63.39	44.64	28.300
1.80	180	.594	56.43	44.10	24.886	67.12	47.27	31.727
1.90	190	.627	59.56	46.55	27.728	70.85	49.89	35.350
2.00	200	.660	62.70	49.00	30.723	74.58	52.52	39.169

Table 4-1   *All dimensions are MM (for British measurements see Table 3-1). (EAN G.S., UPC SSM 1990)*

4. EAN-13 Format and Dimensions

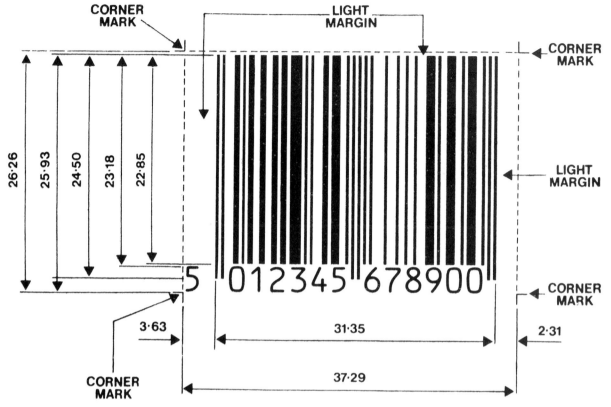

Figure 4-5   *All dimensions are MM. (EAN G.S.)*

5. How to Obtain an EAN Code

The EAN appoints one member organization in each country, responsible for local assignment and control of bar coding. Manufacturer's must join their local EAN agency to obtain their country and manufacturer's ID numbers. Item numbers are assigned by each manufacturer in some countries and by the EAN agency in others. If unable to find a local EAN agency, you can contact EAN headquarters in Brussels, Belgium, or UCC in Dayton, Ohio, United States.

6. A, B & C Character Encoding

There are three ways to encode characters located in positions 1 to 12, these are A, B and C, depending on the bar width of each character, and whether there are an odd number of modules (3 or 5), or an even number (2 or 4) and depending on whether the first and last module (of the seven forming each character) are a space and a bar, or a bar and a space, respectively:

A character:
    Has two bars composed of 3 or 5 modules (odd number)
    First left module: a space
    Last right module: a bar
    Located to the left of the center guard bar in positions 7 through 12, together with B characters.

B character:
    Has two bars composed of 2 or 4 modules (even number)
    First left module: a space
    Last right module: a bar
    Located to the left of the center guard bar in positions 7 through 12, together with A characters.

C character:
    Has two bars composed of 2 or 4 modules (even number)
    First left module: a bar
    Last right module: a space
    Located to the right of the center guard bar in positions 1 through 6.

A and B characters are located to the left of central center guard bar, C characters are located to the right.

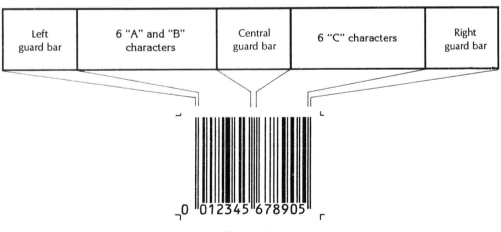

Figure 4-6

53

7. Encoding EAN-13: A, B & C Characters

characters:	A	B	C
number	ODD, 3 or 5 modules	EVEN, 2 or 4 modules	EVEN, 2 or 4 modules
0	0 0 0 1 1 0 1	0 1 0 0 1 1 1	1 1 1 0 0 1 0
1	0 0 1 1 0 0 1	0 1 1 0 0 1 1	1 1 0 0 1 1 0
2	0 0 1 0 0 1 1	0 0 1 1 0 1 1	1 1 0 1 1 0 0
3	0 1 1 1 1 0 1	0 1 0 0 0 0 1	1 0 0 0 0 1 0
4	0 1 0 0 0 1 1	0 0 1 1 1 0 1	1 0 1 1 1 0 0
5	0 1 1 0 0 0 1	0 1 1 1 0 0 1	1 0 0 1 1 1 0
6	0 1 0 1 1 1 1	0 0 0 0 1 0 1	1 0 1 0 0 0 0
7	0 1 1 1 0 1 1	0 0 1 0 0 0 1	1 0 0 0 1 0 0
8	0 1 1 0 1 1 1	0 0 0 1 0 0 1	1 0 0 1 0 0 0
9	0 0 0 1 0 1 1	0 0 1 0 1 1 1	1 1 1 0 1 0 0

Figure 4-7   *Each black module = 1, each white module = 0*

8. EAN Symbol Nominal Dimensions and Encoding Chart

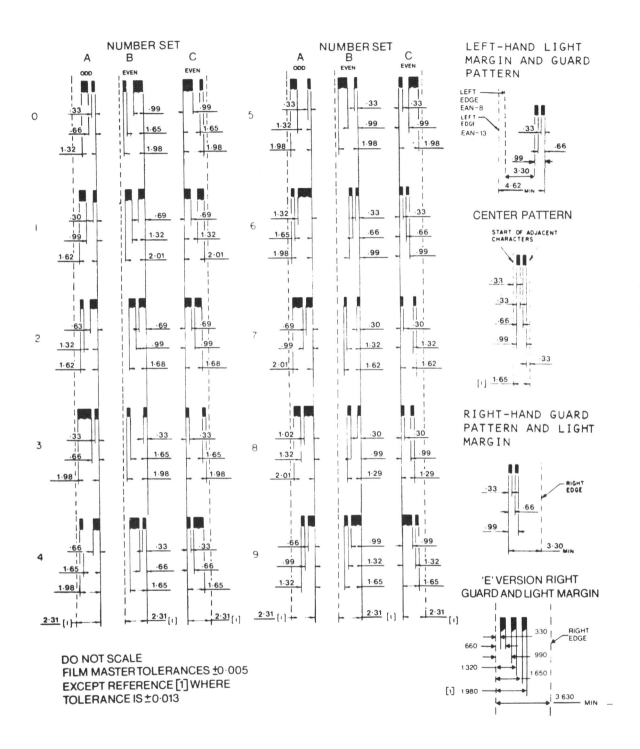

Figure 4-8 *All dimensions are MM. (EAN G.S.)* 55

9. Character 13 Calculation

Sequences of A/B characters located on positions 7 through 12 will determine the character on position 13 (which belongs to the first country ID number or flag). This character will not be symbolized by bars or spaces like the others and is usually printed on the left quiet zone.

Position # 13	12	11	10	9	8	7
Defined number	A & B Character sequence required					
0	A	A	A	A	A	A
1	A	A	B	A	B	B
2	A	A	B	B	A	B
3	A	A	B	B	B	A
4	A	B	A	A	B	B
5	A	B	B	A	A	B
6	A	B	B	B	A	A
7	A	B	A	B	A	B
8	A	B	A	B	B	A
9	A	B	B	A	B	A

Table 4-2   *EAN A/B character sequence chart (EAN G.S.)*

   Therefore, when all characters in position 7 through 12 are A-type, the number "0" is defined for the character in position 13, and this is the specific case in which a European system will read an American UPC symbol (UPC only has characters A and C).
   This situation defines the compatibility of the EAN (European) with the UPC system (American), as European scanners can read a UPC code containing only A-type characters on positions 7-12 and will assign the number "0" to position 13, which does not exist in the UPC system.
   The reverse is not always possible, as much American and Canadian equipment programmed in UPC 12 digit systems only, is unable to understand the B-type character and cannot generate the 13 flag digit needed to read the EAN code. UCC and EAN expect to have both systems fully compatible within a few years.

10. Check Digit calculation

Unlike the others, the check digit located in position 1 is not assigned to an item but is the result of a calculation involving numbers located in positions 2 through 13. This is designed to prevent reading errors and wrong scanning usually caused by defective symbol design or printing, as unassigned numbers could be read. It also features a "self-checking" operation to detect self-calculation errors. Check digit calculation method is the same for UPC-A and EAN-13 codes, as follows:

(a) Multiply each character in an odd position X 1.
Multiply each character in an even position X 3.
(b) Add the 12 numbers obtained, getting a value known as *products sum (PS)*.
(c) Divide the products sum by 10 (a constant), obtaining a quotient (C) and a balance (B);
PS/10 = C and B.
(d) Deduct the balance from 10, obtaining the check digit (CD) to be placed on position 1
10 - B = CD.

- Check digit (CD) calculation using EAN-13 code, number 0012345678905

	Position #	13	12	11	10	9	8	7	6	5	4	3	2	1
	value:	0	0	1	2	3	4	5	6	7	8	9	0	?
(a)	×			1	3	1	3	1	3	1	3	1	3	
	=													
(b)	PS	0 +	0 +	1 +	6 +	3 +	12 +	5 +	18 +	7 +	24 +	'9 +	0	

(b) PS = 85

(c) $\dfrac{PS}{10}$ = C and B   so, $\dfrac{85}{10}$ = 8 and B = 5

(d) 10 − B = CD   so, 10 − 5 = $\boxed{5}$

so Check Digit CD = 5 and we can place it in position # 1

Table 4-3

- Self-checking

Self check of the CD character just calculated is performed using exactly the same procedure as above, introducing the following changes:

(a)   The value in position 1 now exists and is the one calculated above, equal to 5; here the multiplying factor is always 1.

(c)   If the quotient results in zero balance (B = 0), the character value calculated above is confirmed and self checked on position 1.

- Check digit (CD) verification example on position 1 using EAN-13 code number 0012345678905

Position #		13	12	11	10	9	8	7	6	5	4	3	2	1
	value:	0	0	1	2	3	4	5	6	7	8	9	0	5 ←
(a)	×													
		1	3	1	3	1	3	1	3	1	3	1	3	1
	=													
(b)	PS	0 +	0 +	1 +	6 +	3 +	12 +	5 +	18 +	7 +	24 +	9 +	0 +	5

(b)   PS   = 90

(c) $\dfrac{PS}{10}$ = C and B   so, $\dfrac{90}{10}$ = 9 and B = $\boxed{0}$ check digit has been verified

Table 4-4

B. EAN Codes Truncation

When packaging offers such a small bar code printable area that it is not possible to include a proper size EAN-13 code, or when substrate characteristics and printing system (printability gauge) does not even allow for the use of the EAN-8 reduced version, the final and last choice is to cut bar length or truncate the code.

Truncating symbols proportionally reduces a scanner's omnidirectional reading possibilities, which means that the product has to be manipulated, rotated and exposed several times at the scanner until properly read, if it is read. This is a waste of time, which is precisely what the code was designed to eliminate. The greater the truncation, the lower the reading chances are.

When a EAN code is truncated, a basic symbol should be taken with as great a magnification factor as possible, avoiding, if possible, truncation of codes with a magnification factor lower than 100 percent. To determine the minimum truncation possible, the following table is used. Truncation 1 can be done in bar length, but this will depend on the magnification factor of the basic code. If insufficient, truncation 2 could be performed, this being the maximum truncation allowed according to each magnification factor. Always consider that the printed code will probably lose the omnidirectional reading capacity when truncated.

1. EAN Truncation Chart

Base EAN Symbol Magnification Factor (MF)	Bar length reduction		
	Reduction 1 (mm) from: to:	Reduction 2 (mm)	Reduction 1 + 2 (mm) from: to:
0.8 - 0.95	No Truncation	No Truncation	No Truncation
1.0	No Truncation	3.8	3.8 3.8
1.1	0.8 0.8	3.8	4.6 4.6
1.2	1.5 1.6	3.8	5.3 5.4
1.3	2.3 2.4	3.8	6.1 6.2
1.4	3.0 3.2	3.8	6.8 7.0
1.5	3.8 4.0	3.8	7.6 7.8
1.6	4.6 4.8	3.8	8.4 8.6
1.7	5.3 5.6	3.8	9.1 9.4
1.8	6.1 6.4	3.8	9.9 10.2
1.9	6.9 7.2	3.8	10.7 11.0
2.0	7.6 8.0	3.8	11.4 11.8

Table 4-5 *Truncation chart does not apply to UPC symbols (EAN G.S.)*

1   1 + 2

0 012345 678905   0 012345 678905   0 012345 678905

MF = 1.2   −1.6 mm   −5.4 mm

Figure 4-9 *Correct EAN Truncation example, MF = 1.2*

C. EAN-8 Code

EAN-8 is the reduced version of the EAN system, used only when packaging size and/or shape leave no printing room available for an EAN-13 code. Although a shorter version of the latter, the EAN-8 code is not just another way of storing the same information, as EAN-8 is a completely independent code only available to the manufacturer if assigned by the local EAN agency, unlike UPC-E, which is simply a reduced version of UPC-A.

Advantages of the EAN-8 systems are:
Smaller, requires less space.
As reliable and readable as EAN-13 (at the same magnification factor).
The use of EAN-8 is recommended over EAN-13 truncation.
Disadvantages of the system:
Limited encoding capacity.
The use of EAN-8 is not free and should be assigned by the local encoding organization.
Obtaining EAN-8 by zero suppressing is no longer used.

EAN has defined the reduced version to permit encoding 10,000 items, assigning three digits per country (the traditional flag), four digits per item, and a check digit to complete the system. Manufacturer's code is excluded. This version has been used since 1987 as a local encoding organization option.

Application for EAN-8 instead of EAN-13 is possible when one of the following situations results from printing trials with the printability gauge:

(a) Minimum adequate EAN-13 MF is larger than 25 percent of the larger panel printed area, or the panel is less than 40 cm² (15.7 sq.inch).
(b) Minimum adequate EAN-13 MF is larger than 12.5 percent of the whole printed area, or the area is less than 80 cm² (31.4 sq.inch).
(c) Available area is cylindrical, and diameter is less than 3 cm (1.2'').

Figure 4-10 *EAN-8 code example, MF =1.0*

1. EAN-8 Characteristics

Characters: A total of 8, numbers only, as follows:
Character. 8,7,6:   Country ID (Flag)
Character. 5,4,3,2:   Item ID
Character. 1:   Check digit

The characters are single-digit numbers (0 1 2 3 4 5 6 7 8 9), 8 of which will form the EAN-8 code (as an example only, code 00123457 is used in this chapter and diagrams).

Eight characters will form the symbol represented and printed as bars and spaces. Each character is represented by 2 bars and 2 spaces alternately arranged, i.e., 4 elements per character; element width and location make the difference between one character and the other.

Character width is fixed, measuring 7 modules (a module is the thinnest element). Therefore, all four elements of a character will have a total width of 7 modules, so each bar and/or space will be 1 module wide minimum and 4 modules wide maximum, thus forming a complex structure code. The above criteria apply only to the eight characters encoded in EAN-8 system, and do not apply to guard bars or quiet zones.

Character location, or positioning in the code, is defined as follows: Position 1 is the first character to the right of code; position 8, is the last character to the left of code. This way of locating characters is only for study purposes and is not part of the codification.

This apparent complexity has the purpose of allowing the computer decoder to identify the code, find the correct direction in bi-directional reading, activate electronic checking mechanisms, avoid errors and prevent wrong scans.

However, these complications should not frighten users as the film masters needed for symbol printing are computer-generated. Handmade design and drawing are useless and unnecessary. Film master computerized photosetting only requires the complete number assigned, size (MF), code type and a few photographic details each printer is be able to obtain.

- Guard bars pattern:
    Left, fixed width: 3 modules, BAR-SPACE-BAR, encoded 101
    Right, fixed width: 3 modules, BAR-SPACE-BAR, encoded 101
    Center, fixed width: 5 modules, SPACE-BAR-SPACE-BAR-SPACE, encoded 01010
    Guard bar standard height: 19.99 mm (0.79''), slightly higher than most bar
- Quiet zones:
    Left: 7 modules wide minimum, encoded 0000000
    Right: 7 modules wide minimum, encoded 0000000
    Lower: 1 module, between bars and character set
- Encoding: continuous, bi-directional
- Character set:
    Location: all 8 characters at the bottom of code
    Characters font type: OCR-B
- Structure: complex
- Printed module, standard, (MF = 1):
    Width: 0.330 mm ± 0.101 mm (0.013'' ± 0.004'')

- Fixed length:
    81 modules between corner marks
    67 modules between right and left guard bar patterns
- Standard density: medium
- Standard size (MF = 1)
    26.73 × 21.64 mm (1.05″ × 0.85″) between corner marks,
    including quiet zones (right, left and lower)
- Symbol height (bar or space):
    18.23 mm (0.72″) except for guard bars
- Magnification factors:
    MF = 0.8 to 2.0 (80 percent to 200 percent)
- Compatibility: can be understood by all EAN scanners and some UPC systems
- Code administration: EAN local organization
- Flag: The flag or country ID number is 3 digits wide. See EAN flags chart, Table: 4-10.

2. EAN-8 Magnification Factors and Dimensions

EAN-8 Magnification Factors		Module width (M)	Between first and last bar			Between Corner Marks		
			width	length	surface	width	length	surface
(MF)	(%)	mm	mm	mm	cm²	mm	mm	cm²
0.80	80	.264	17.69	15.90	2.813	21.38	17.31	3.702
0.85	85	.281	18.79	16.90	3.176	22.72	18.39	4.179
0.90	90	.297	19.90	17.89	3.560	24.06	19.48	4.685
0.95	95	.314	21.00	18.89	3.967	25.39	20.56	5.220
1.00	100	.330	22.11	19.88	4.395	26.73	21.64	5.784
1.10	110	.363	24.32	21.87	6.329	29.40	23.80	6.999
1.20	120	.396	26.53	23.86	5,319	32.08	25.97	8.329
1.30	130	.429	28.74	25.84	7.428	34.75	28.13	9.776
1.40	140	.462	30.95	27.83	8.615	37.42	30.30	11.337
1.50	150	.495	33.16	29.82	9.890	40.09	32.46	13.015
1.60	160	.528	35.38	31.81	11.252	42.77	34.62	14.808
1.70	170	.561	37.59	33.80	12.703	45.44	36.79	16.717
1.80	180	.594	39.80	35.78	14.241	48.11	38.95	18.741
1.90	190	.627	42.01	37.77	15.868	50.79	41.12	20.882
2.00	200	.660	44.22	39.76	17.582	53.46	43.28	23.137

Table 4-6   *All dimensions are MM. (EAN G.S.)*

3. EAN-8 Format and Dimensions

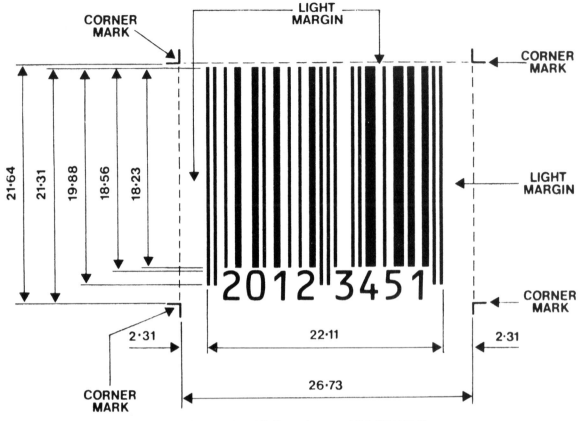

Figure 4-11 All dimensions are MM. (*EAN G.S.*)

- How to Obtain an EAN code

The EAN appoints one member organization in each country, who is responsible for assigning and controlling local bar coding. Manufacturers must join their local EAN agency to obtain their ID numbers. If unable to find a local EAN agency, you can contact EAN headquarters in Brussels, Belgium, or UCC in Dayton, Ohio, United States.

4. A & C Characters Encodation

There are two ways to encode characters located in positions 1 to 8; these are A and C, depending on the bar width of each character, whether there is an odd number of modules (3 or 5), or an even number (2 or 4), and whether the first and last module (of the seven forming each character) are a space and a bar, or a bar and a space, respectively:

A character:
 Has two bars composed of 3 or 5 modules (odd number)
 First left module: a space
 Last right module: a bar
 Located to the left of the center guard bar on positions 5 through 8

C character:
 Has two bars composed of 2 or 4 modules (even number)
 First left module: a bar
 Last right module: a space
 Located to the right of the center guard bar on positions 1 through 4

A characters are located to the left of central center guard bar, while C characters are located to the right.

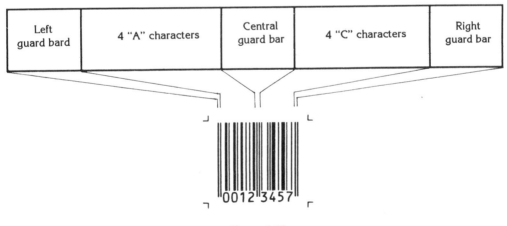

Figure 4-12

5. A & C Character EAN-8 Encoding

The EAN-8 Nominal Dimensions and Encoding Chart is included in Figure 4-8.

Characters:	A	C
Number	ODD, 3 or 5 modules	EVEN, 2 or 4 modules
0	0 0 0 1 1 0 1	1 1 1 0 0 1 0
1	0 0 1 1 0 0 1	1 1 0 0 1 1 0
2	0 0 1 0 0 1 1	1 1 0 1 1 0 0
3	0 1 1 1 1 0 1	1 0 0 0 0 1 0
4	0 1 0 0 0 1 1	1 0 1 1 1 0 0
5	0 1 1 0 0 0 1	1 0 0 1 1 1 0
6	0 1 0 1 1 1 1	1 0 1 0 0 0 0
7	0 1 1 1 0 1 1	1 0 0 0 1 0 0
8	0 1 1 0 1 1 1	1 0 0 1 0 0 0
9	0 0 0 1 0 1 1	1 1 1 0 1 0 0

Figure 4-13   *Each black module =1, each white module =0*

6. Check Digit calculation

Unlike the others, the check digit is located in position 1 and is not assigned to an item but is the result of a calculation involving numbers located in positions 2 through 8. This is designed to prevent reading errors and wrong scanning usually caused by defective symbol design or printing, as unassigned numbers could be read. The check digit also features a self-checking operation to detect self-calculation errors. The check digit calculation method is the same for UPC-A and EAN codes, as follows:

(a) Multiply each character in an odd position X 1.
   Multiply each character in an even position X 3.
(b) Add the 7 numbers obtained,
   obtaining a value known as products sum (PS).
(c) Divide the products sum by 10 (a constant),
   obtaining a quotient (C) and a balance (B);
       SP/10 = C and B.
(d) Deduct the balance from 10,
   obtaining the check digit value (CD) to be placed in code position 1;
       10 - B = CD.

- Check digit (CD) calculation using EAN-8 code, number 00123457

Position #		8	7	6	5	4	3	2	1
(a)	value: X	0 3	0 1	1 3	2 1	3 3	4 1	5 3	?
(b)	PS	0 +	0 +	3 +	2 +	9 +	4 +	15	

(b)	PS = 33		
(c)	$\dfrac{PS}{10}$ = C and B	so, $\dfrac{33}{10}$ = 3 and B = 3	
(d)	10 − B = CD	so, 10 − 3 =	7
(d)	so Check Digit CD = 7 and we can place it in position # 1		

Table 4-7

- Self check

  Self check of a CD character just calculated is performed using exactly the same procedure as above, with the following changes:

  (a)   The value in position 1 now exists and is the one calculated above, equal to 7; here the multiplying factor is always 1.

  (c)   If the quotient results in zero balance (B = 0), the character value calculated above is confirmed and self checked in position 1.

- Check digit (CD) verification example in position 1 using EAN-8 code number 00123457

Position #			8	7	6	5	4	3	2	1
	value:		0	0	1	2	3	4	5	7
(a)	× =		3	1	3	1	3	1	3	1
(b)	PS				0 + 0 + 3 + 2 + 9 + 4 + 15 + 7					

(b)	PS	= 40			
(c)	$\dfrac{PS}{10}$ = C and B		so,	$\dfrac{40}{10}$ = 4 and B =	0
	check digit has been verified				

Table 4-8

Figure 4-14   *EAN-8 code MF = 2*

D. Book and Magazine International Codification

1. Books (ISBN)

Numbering of books published all over the world is done by the International Standard Book Numbering System (ISBN) through its representative in each country.

Thanks to agreements between EAN, UCC and ISBN, the EAN and UPC codes can be used to encode any book's ISBN numbering anywhere in the world.

EAN has assigned flags 978 and 979 to this international agency. Thus, the ISBN symbol is composed as follows:

ISBN Flag (978 or 979):	3 digits
ISBN book number:	9 digits (eliminating original check digit)
ISBN check digit:	1 digit recalculated for this code

For example, if the ISBN number for a book is 950431979-3, the EAN correct code will be 978-950431979-5.

Figure 4-15   *ISBN example, MF =1*

2. Magazines (ISSN)

Numbering of magazines and serial publications all over the world is done by the International Standard Serial Numbering System (ISSN). Thanks to an agreement similar to the one described above it is possible to encode ISSN numbering under EAN/UPC anywhere in the world, for which EAN has assigned flag 977.

Thus, the ISSN symbol is composed as follows:

ISSN flag (977):	3 digits
Publication's ISSN number:	7 digits (eliminating original check digit)
Extra ISSN digits:	2 digits
ISSN new check digit:	1 digit recalculated for this code

Figure 4-16 *ISSN examples* (Courtesy of Stork Graphics)

E. In-Store UPC/EAN Codification

EAN and UPC codes foresee the need for local assignment of codes for limited circulation (in-store or locally assigned codes: LAC) within a store or an area where the code does not interfere with general codification used elsewhere. Domestic LAC codes can be repeatedly used, but they must differ from general EAN or UPC codes.

Those users generating this type of code should be responsible for guaranteeing that its circulation will actually be "local and limited" to the store or supermarket and to products not coded under the general system in use, such as random weight items such as meat, fruit, vegetables, and special sales of uncoded items.

- Locally Assigned Codes

  EAN and UPC have reserved flag 2 for local encoding, and the various remaining digits available should be determined by the user (prices can also be encoded, obviously, in low-inflation countries only, and a second check digit can be added for the price if necessary). For this reason no country flag starts with 2.

- In-store encoding quick reference

EAN 13	2	X	X	X	X	X	X	X	X	X	X	C
UPC A	Z	2	X	X	X	X	X	X	X	X	X	C
EAN 8	Z	Z	Z	Z	Z	2	X	X	X	X	X	C
UPC E	Z	0	0	B	B	B	B	0	0	0	E	C

2:	In store encoding, flag # 2
0:	0 number
X:	Digits locally assigned by the user
C:	Check digit
B:	Any value between 1000 and 7999 (for UPC-E only)
E:	Any value between 5 and 9 (for UPC-E only)
Z:	Nonexistent 0 number (will only be scanned by EAN-13)

Table 4-9

F. EAN Flags Chart

Flag prefix digits assignment and estimated number of members (EAN Jan. 1993)

Country	Flag ID	Organization Responsible	City	Members (Jan. 93)
Argentina	779	CODIGO	Buenos Aires	2,990
Australia	93	APNA	Victoria	5,249
Austria	90 , 91	EAN Austria	Vienna	2,821
Belgium	54	ICODIF	Brussels	1,729
Books	978 , 979	(ISBN)	—	—
Brazil	789	ABAC	Sao Paulo	1,606
Bulgaria	380	CCI Bulgaria		
Canada	00 to 09	UCC	Dayton, USA	15,000
Chile	780	DEPCO	Santiago	1,107
China	690	ANCC		
Colombia	770	IAC	Bogota	420
Coupons	98 , 99	Coupon codes only	—	—
Cuba	850	CC of Cuba	Havana	6
Croatia	385	CRO-EAN	Zagreb	113
Cyprus	529	CYPRUS CCI	Nicosia	344
Czechoslovakia	859	CSS EAN	Prague	440
Denmark	57	DVA	Hellerup	2,300
Ecuador	786	ECOP		
Finland	64	CCCF	Helsinki	11
France	30 to 37	GENCOD	Paris	15,000
Germany	400 to 440	CCG	Cologne	25,000
Greece	520	HELLCAN	Athens	854
Guat., C. America	740 to 745	ICCC	Guatemala	180
Hong Kong	489	HKANA	Hong Kong	902
Hungary	599	CAOS/MGK	Budapest	845
Iceland	569	I.EAN CMT	Reykjavik	294
Ireland	539	ANAI		756
In - Store	20 to 29	In - Store numbers	—	—
Israel	729	ICA	Tel Aviv	1,147
Italy	80 to 83	INDICOD	Milan	14,327
Japan	45 ,49	DCC	Tokyo	67,898
Luxembourg	54	ICODIF	Brussels	1,996
Magazines	977	(ISSN)	—	—
Malaysia	955	MANC	Kuala Lumpur	341
Malta	535	MANA	Malta	
Mexico	750	AMECOP	Mexico City	4,650
Netherlands	87	STICHTING UAC	Amstelveen	1,981
New Zealand	94	NZPNA	Wellington	2,581
Norway	70	NORSK VAREK F.	Oslo	5
Peru	775	APC	Lima	194
Poland	590	BCC	Poznan	515
Portugal	560	CODIPOR	Lisbon	2,010
Russian Federation	460 to 469	UNISCAN		74
Singapore	888	SANC	Singapore	741
South Africa	600 , 601	SAANA	Johannesburg	3,334
South Korea	880	KANC	Seoul	689
Spain	84	AECOC	Barcelona	8,433
Sweden	73	S.EAN CMT	Stockholm	2,426
Switzerland	76	SACV	Vevey	1,317
Slovenia	383	SANA	Ljubljana	204
Taiwan/China	471	ANC	Taiwan	3,112

Flag prefix digits assignment and estimated number of members (EAN Jan. 1993) *(Cont.)*

Country	Flag ID	Responsible Organization	City	Members (Jan. 93)
Thailand	885	THAI PNA		115
Tunisia	619	TUNICODE	Bangkok	
Turkey	869	UCCET	Ankara	440
UK	50	ANA	London	8,769
Uruguay	773	CUNA	Montevideo	58
U.S.A.	00 to 09	UCC	Dayton	75,000
Venezuela	759	CIP	Caracas	79
Yugoslavia	860	YANA		

| Estimated total | Jan. 31, 1993 | | | 280,403 |

Table 4-10 (EAN Af. 1993)

G. Code Interleaved 2 of 5 (ITF)

Code "2 of 5" was developed in 1968 and later proposed as "interleaved 2 of 5" in 1972. It is used for product and container identification for warehousing and distribution. This is a general purpose code used especially for industrial applications.

USS I 2/5 is the specification of code I 2/5; USS (Uniform Symbol Specification) was developed by "AIM Technical Symbology Committee" established in 1983 by AIM Inc. This committee is formed by staff members of several corporations participating in AIM, elected by their members.

Introduction: USS I 2/5 (code interleaved 2 of 5) is a bar code symbology with a set of number characters and different start and stop characters. The name "Interleaved 2 of 5" derives from the method used to encode pairs of characters. In the symbol, *two* characters are paired using bars to represent the first character and spaces to represent the second. Each character (0 through 9) consists of two wide and two narrow elements, for a total of *five* bars or spaces (*two of five*).

- Characteristics:
    - Encodable characters: numbers only
    - Code type: continuous
    - Symbol length: variable, it encodes any even number of digits
    - Decoding: bi-directional
    - Number of check digits required: none
    - Character self check: yes
    - Minimum nominal module: 0.0075'' (0.191 mm)
    - Maximum density: 18 char/inch (7.1 char/cm)
    - Check digit: optional
    - Additional characteristic: unique start and stop character

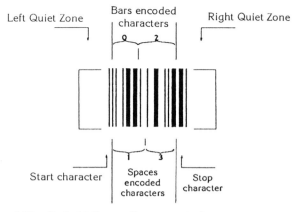

Figure 4-17 *Code I 2/5 encoding example for message: 0 1 2 3*

H. Code 39 (3 of 9)

This code was created in 1974 and adopted in the United States in 1982 by the Department of Defense (LOGMARS), Service Administration (GSA) and the Automotive Industry (AIAG). The SCS (Shipping Container System) later renamed the code, and Uniform Container Symbol was adopted for storage and distribution; the code was originally designed to be printed on corrugated cardboard.

USS-39 is the specification of code 39; USS (Uniform Symbol Specification) was developed by the AIM Technical Symbology Committee established in 1983 by AIM Inc. This committee is formed by staff members from several corporations participating in AIM, elected by their members.

Introduction: USS-39 (code 39) is a bar code symbology which includes a complete series of alphanumeric characters, unique start and stop characters, and seven special characters. The name "39" comes from its code structure that consists of *three* wide elements out of *nine*. These nine elements are composed of five bars and four spaces each.

- Characteristics:
    Alphanumeric characters (all of them) and the seven special characters -.Space $/+%
    One start/stop character: *
    Code type: discrete
    Symbol length: variable
    Decodification: bi-directional
    Character self check: yes
    Number of check digits: none
    Minimal nominal module: 0.0075″ (0.191 mm)
    Maximum density: 9.8 char/inch (3.7 char/cm)

— Each symbol consists of:
    a - Left quiet zone
    b - Start character
    c - One or more data characters
    d - Stop character
    e - Right quiet zone

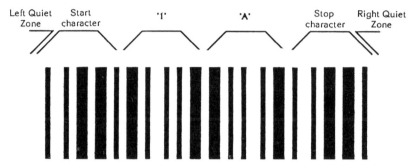

Figure 4-18   *Code 39 encoding example for message: 1 A*

I. Other Bar Code Systems and Specifications

AGES
AS 6/10
B code
BCD (binary code decimal) modified BCA
BP (binary periodic)
CODABAR / 2 of 7 / USD-4 / Ames / USS-Codabar
Code 11 / USD-8
+ Code 39 / 3 of 9 / USD-3/2 / USS-39 / (ANSI, LOGMARS, HIBC, AIAG Automotive)
Code 49 ultra high density
Code 93 / USD-7 / USS-93
Code 128 / USD-6 / USS-128
Code PDF 417
Delta Distance A/B
Distribution symbol
* EAN (European article numbering)
Frequency code
IAN (Int'l. Art. Numbering) / WPC (world product code)
+ Interleaved 2 of 5 / USD 1 / I 2/5 / USS-I 2 5 / (ITF, STRAIGHT, IATA)
Nor and / F2F
Octal decimal code
Plessey / PWM (pulse width modulation) / Anker / MSI

RTC
SCOPE code
* UPC (Universal Product Code) A/E

* Codes mainly for commercial use (at POS) extensively described in this book.
+ Codes mainly for industrial use (and other applications) briefly described in this book.

5

Shipping Container Code

Encodable commercial products are classified by EAN as *consumer units* or *despatch units*, according to the following international criteria:

A. Consumer Units

The Consumer Unit of a product is the item to be sold to final consumers at a point of sale (e.g., checkout counter at a department store or supermarket). Consumer units or items are encoded under EAN or UPC systems, using UPC-A, UPC-E, EAN-13 or EAN-8 symbologies, as explained in previous chapters. Consumer units can be classified in two groups as follows:

1. Basic Consumer Unit

The basic consumer unit is an item that cannot be partitioned or divided in order to be sold at the point of sale

2. Multipack

Multipack is an item composed of many basic consumer units that cannot be partitioned or divided in order to be sold at the point of sale

B. Despatch Units

Every standard and stable group of consumer units that *will not* be sold to final consumers at a point of sale is a "despatch unit". These units are traditionally the containers and boxes that manufacturers send to retail outlets (e.g., supermarkets) containing consumer units.

These cases or boxes are used for shipping, despatch, delivery and transportation. Despatch units are internationally encoded by the Interleaved 2 of 5 or "ITF" bar code system, using ITF-14 symbology, though ITF-16 or EAN-13 may also be used locally within each country, following local EAN/UPC encoding association criteria.

ITF symbology is explained in Section 4-G.

Figure 5-1 *ITF-14 code example, MF = 1*

C. ITF-14 Encoding

A despatch unit number (DUN) is based on the number of consumer units (CUN) it contains; the number consists of a total of 14 digits, divided into the three following groups:

A   Logistical variant (LV): 1 digit
    Indicates number of consumer units the despatch unit contains
    LV = 9 indicates *variable amount and add-on* code presence
B   Consumer unit or EAN/UPC code: 12 digits (the EAN-13/UPC-A code, *without* check digit)
    F: Flag or country ID: 2/3 digits
    M: Manufacturer identification: 5/4 digits
    I: Item identification: 5 digits
C   New check digit: 1 digit
    calculated from the 13 characters from items A and B above

- ITF 14 Structure

A Logistical variant (1 digit)	B EAN 13 - Consumer unit without original check digit (12 digits)	C Check digit (1 digit)
LV	F F F M M M M I I I I I	C

Table 5-1

D. ITF Magnification Factor (MF) and Contrast (PCS)

All ITF symbol specifications refer to *standard* dimension when MF = 1 (Figure 5-1); dimensions may vary between 0.625 and 1.2 of standard size (MF = 0.625 to 1.2). Specifications are effective both for main and add-on symbols.

The minimum print contrast signal (PCS) value permitted is 75 percent for ITF symbology, higher than the minimum PCS of 65 percent required for EAN codes. The spaces' reflectance should be higher than 25 percent for ITF symbols.

E. ITF Additional Encoding

There are two additional encodings:

1. Supplementary Encoding: All bar code printed data with *no* logical relation to main symbol, e.g., packaging date, lot or plant number, expiration date. Alphanumeric code 128 is used, and UPC, EAN or ITF symbols *should not* be printed for this purpose.

2. Variable Quantities (Add-on): Refers to items whose content price may continuously vary according to the quantity, rather than a fixed number of consumer units, and is directly related to the main symbol. Quantity contained may be an area, volume, weight, length, units, etc.:

- *Contents* are identified by EAN/UPC or ITF codes.
- *Quantities* are identified by the add-on code (*addendum*)

- The add-on encoding consists of a symbol printed to the right, following the main symbol to which it necessarily belongs. For despatch units, an ITF-6 symbol is used (6 character, Interleaved 2 of 5).

Specifications and regulations for encoding despatch units are supplied by local EAN or UPC organizations in each country, although all countries observe international regulations for foreign trade.

6

Bar Code Design
and Printing

We recommend the appointment of only one person or team responsible for bar coding, because the system directly or indirectly involves many different areas. Bar codes should be handled by a person or group specifically trained and capable of mastering this new language to provide maximum benefit to the organization within domestic and international specifications.

A. Packaging Design

Companies new to this technology (who still do not have an expert in automatic ID systems) are often faced with complaints from designers about having to include these "unesthetic bars with their aggressive size, colors and appearance" in their package design. Such frustration may lead to unnecessary and undesired design errors, jeopardizing the aim of the bar code: *To be successfully scanned at the point of sale, automatically and repeatedly.* These are the usual consequences of inexpert bar code personnel:

- Reducing the code even below minimum specifications
- Truncating the code when less drastic solutions are available
- Mistaking one symbol for another or choosing the wrong code number
- Deciding on a code location on the package only because it is less visible
- Choosing symbol colors only for artistic reasons
- Ignoring base film, packaging and contents in art design
- Ignoring the printing system to be used
- Considering a job to be over when actually only the basic design has been finished
- Failing to verify estimated design performance on the final package

These conditions might cause symbol scanning failure and product return or complaints and could be avoided by providing designers, art departments, photographers, cylinder engraving and plate sections with all the information needed to understand the bar code language and the different stages of the process they are involved in.

Professional designers should have all the available elements (manuals, specifications, instruments) helpful in providing designs according to international requirements and specifications. UPC/EAN standards plus the appointed bar code responsible will supply these guidelines so that only permitted tools are used to design a perfectly printed code that can be scanned successfully worldwide.

This chapter will help designers avoid the errors listed above and provide them with information needed to optimize results.

B. Bar Code Printing Options

The bar code symbol of a consumer unit must be on the product, either printed by the packaging manufacturer on the package, or on a label printed on-site or off-site.

These are the basic options and their usual printing systems:

- On-the-package, Printings with standard web widths from 24'' (600 mm) to 63'' (1600 mm), are explained in Chapter 7. The printing systems are:
    - Gravure (rotogravure)
    - Flexography, rubber or photopolymer
    - Offset, wet or dry
    - Silk screen, flat or rotary
    - Tampography

- Labels with standard web widths from 2'' (50 mm) to 24'' (600 mm), are specifically explained in Chapter 9. Their printing systems include:
    - Thermal, transfer or direct
    - Impact
    - Laser
    - Ink Jet
    - Dot matrix
    - Electrostatic and ion deposition
    - Narrow web flexo, rubber or photopolymer
    - Silkscreen

C. The Photographic Process

Although the code is usually part of the printing of a product package, the art, photography, quality control and printing processes involved are far more complex than standard art design. Precision symbol requirements are as critical as for color photography, separation or register control in gravure printing. For designers, most potential defects are not always visible. Some only appear while printing the package, and others are found too late at the point of sale when the scanner is unable to read. This usually happens because of a mistake or series of different errors in the process that starts with the film master and ends when the product is scanned. Every step, at every stage, should be carried out very carefully according to standards and specifications. Improvisation should be avoided at any cost, and adequate advice should be sought when needed.

1. Film Master

The film master is a highly accurate, original photographic film, automatically generated by computerized photosetting equipment programmed with the bar code system program to be used. Film master suppliers use sophisticated high-tech electronic equipment for the production, verification and quality control of film masters. Any film master received at the printer's art design or photo engraving department should be carefully checked, since film masters are the first possible source of future problems.

The film master includes the whole set of symbol elements (bars, spaces, quiet zones, interpretation lines), even frame marks. Many film master suppliers today add some extra elements such as a printability gauge, code type, magnification factor, density, customer and product name and number, production date.

If not furnished by the supplier, manufacturer and item names should be printed on the film master after carefully verifying the IDs to prevent encoding the wrong item. A list of customers and items and their bar code numbers should be kept as a constant control and reference.

Film masters must be ordered by the printer, not by the manufacturer, since it is the printer who can supply the BWR, MF, emulsion position, color and other data, after performing the print gain tests with the printability gauge. The necessary film master copies should also be requested when needed, to avoid film master duplication when not absolutely necessary.

The standard magnification factor usually asked for when ordering a film master is MF = 1, although the exact MF to be printed should always be requested on the film master. Careful reduction is possible within allowed magnification limits but is not recommended. The film master should not be drawn or retouched as the accuracy and tolerance of photoelectronic devices can never be obtained by hand.

International pricing for a good film master varies from US$ 5 to US$ 20 per unit, depending on the country, type, quality, quantity and supplier. As film masters are original photographic films, they must be carefully handled and stored. They can be ordered either positive or negative, with emulsion up or down. If flexography is the printing system, this should also be indicated.

Positive, Emulsion Down, MF = 1   Negative, Emulsion Down, MF = 1   Positive, Emulsion Up, MF = 1   Negative, Emulsion Up, MF = 1

Figure 6-1 *These are the 4 emulsion choices available for film masters, seen from emulsion side*

2. Film Master Tolerances

High-precision levels required by UPC/EAN standards oblige film master suppliers to guarantee the lowest possible tolerances. The reliability of their film masters compared with the competition depends on this. Guaranteed specifications should always be indicated.

Standard tolerance for UPC or EAN film masters is ± 0.0002'' (± 0.005 mm) for a standard symbol (MF = 1) with a 0.013'' (0.33 mm) wide module, representing a maximum permitted width error of only ± 1.5 percent in the film master module width. This obviously does not apply to printed symbols where the permitted width error is ± 30.8 percent.

Figure 6-2 *Reduced film master examples (Courtesy of Stork Graphics)*

3. From Film Master to Plate or Cylinder Engraving

Depending on the process, the original art design and symbol film master will merge into the original films to be used in plate or cylinder engraving by the packaging printer. Each color to be printed on the package will have a corresponding photographic film or color separation.

All necessary copying or reduction processes involving film master should be done with emulsions facing each other, taking special care to avoid any type of distortion as far as possible. Once the original films have been obtained, they must be analyzed and checked using high precision verification equipment before plate or cylinder engraving (this test will only confirm whether bar widths conform to specifications and whether the code is correct; no data concerning color, contrast or reflectance will be gathered, as the film will only show black bars on a transparent background). Code and numbers on the original film should also be checked to ensure that they correspond to the correct item.

There is no such thing as a "universal film master". Manufacturers will not have a single film master suitable to all printers. Each printer will buy a film master fitting the specific printing machine, base film and conditions of each printing job according to test results obtained with the printability gauge and design conditions. Printers should always acquire their film masters specifying BWR, MF and other parameters for to each job. This is why a correctly specified film master can only be used by a specific printing shop on a given machine under given conditions on specific substrates.

A film master requested for flexography is different from one specified for another system, although those requested for rotogravure or lithograph/offset can be used in any system other than flexography.

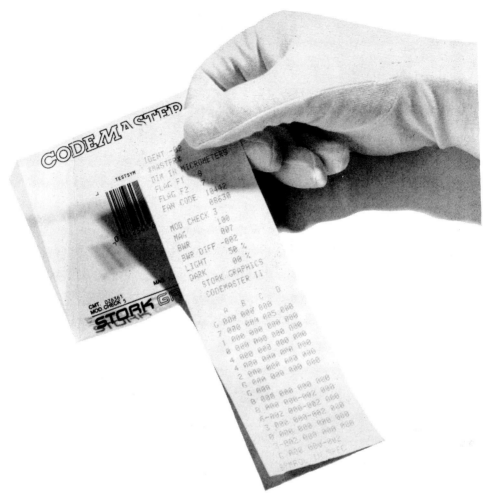

Figure 6-3 *Film master and verification slip examples (Courtesy of Stork Graphics)*

The film master purchased should always come with the corresponding verification slip issued by a film master verification equipment, as shown here, to certify it is within specifications. The printer should never try to use lab verification equipment on film masters, since printed symbol verifiers' module tolerance is ± 30.8 percent (MF = 1), while on film masters it is only ± 1.5 percent. However, a film master lab verifier is more than adequate for a printed symbol but is much more expensive.

The following three examples of different film masters (MF = 1, BWR = 0) show the lower end of a guard bar pattern enlarged over 660 times. On each picture all three elements (bar/space/bar) should be one module wide each (0.013″ = 0.33 mm).

1 - This high-quality film master is below the minimum specifications (± 0.0002″ or ± 0.005 mm), produced by computerized photosetting equipment on a 110 micron photographic film. High-priced, suitable to any printing system and code size.

2 - This medium-quality film master was produced by a laser photoprinter on a 80 micron photographic film. Medium-to low-priced, possibly suitable for offset, flexography or gravure if MF is higher than 0.95.

It does not conform to specifications because the space module was invaded by the bars, whose edge is not perfectly straight.

3 - This is a no-quality master, not conforming to specifications. Produced by a laser printer on non-photographic paper. Again the space was strongly invaded by the bars, which also show a very irregular edge.

This could only serve for low-quality codes when MF is higher than 1.8. Not suitable for silkscreen and low-precision printing techniques in general, regardless of magnification factor.

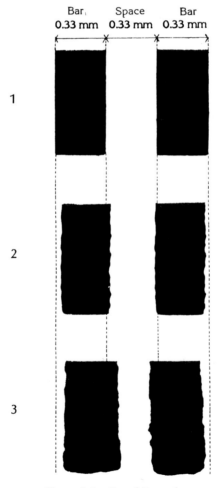

Figure 6-4 *Guard Bar enlargement*

Conclusion: CODE OUTSIDE SPECIFICATION - REJECTED

These four words will frighten any manager, particularly when large productions are rejected by the customer or even worse by a supermarket or foreign customer. Printers should consider whether it pays to run such a risk to save a few dollars on film masters. Obviously, *the answer is no.* Printers should buy only high-quality film masters according to their own MF and BWR needs, emulsion type and form, they should not accept generic masters supplied by third parties, papers or photocopies.

4. Acceptance Tests

Printers often send acceptance tests to their customers for color, text, size and register confirmation. These are matched against standard quality control patterns and then accepted or rejected. Test sets might include one sheet per color, printed only in that color on white paper, plus one sheet where all colors have been printed together and properly registered. Test sets are usually obtained with a proofmaker, not the printing machine.

Manufacturers should never make the mistake of taking acceptance tests or the printed product to a supermarket to scan them, thinking that, if recognized, the code is correct. These readings are meaningless. In the same way they accept tests for their text and color, they should verify that the printed symbol is correct and within specifications. To do this, the analyzing equipment in their quality control department should be similar to that used by their print supplier. Code symbol in the color test sample has a white (paper) background, probably different from the actual background of the final package, and bar color might also be different. The full color sheet taken at the proof maker is subject to similar alterations, plus the stretching and shrinking likely to occur during the printing process, as well as changes in size, contrast and texture resulting from the use of different packaging materials. Certain verifications on test samples such as reflectance and contrast will not be permitted, but others like MF, bar width, code location on package, color selection and code number verification should be performed.

Therefore, code printed in acceptance tests should be carefully checked using appropriate lab instruments. Tests will be conducted in the company's own quality control department or outside service but never at a point of sale. Once again, printed symbol and code numbers should be checked to find out if they really correspond to the item.

D. Code location on the package

The awareness of concept and purpose will aid good decision making. The ultimate purpose of a bar code is *to be successfully scanned at the point of sale, automatically and repeatedly.* This is what usually happens:

a - The consumer selects an item from the shelf.
b - The items are brought in a cart to the supermarket checkout counter (point of sale).
c - The items are placed one by one on the checkout table.
d - The employee takes items one by one and slides them over the scanner.

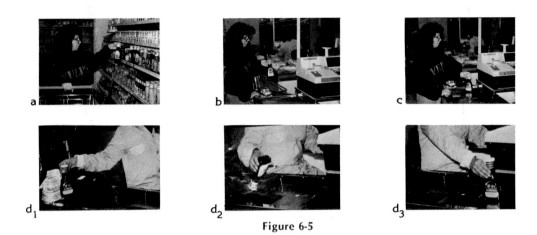

Figure 6-5

There is a natural tendency for consumers to place the purchased items on the checkout table (c) in an upright position on their NATURAL BASE, exactly as they were found on the shelf (a), so this is most likely the item position when the employee slides it over the scanner. Therefore the bar code should be as close as possible to the natural base of each item so the employee will not waste time manipulating items over the scanner.

We define our generic package as formed by a front panel, a rear panel, side panels, a top and a natural base. The package is standing on its natural base, and the front panel is the one facing consumers when arranged on a supermarket shelf.

These concepts apply when designing the code on a package; this may help designers, printers, manufacturers and distributors working on preprinted packages or labeled items.

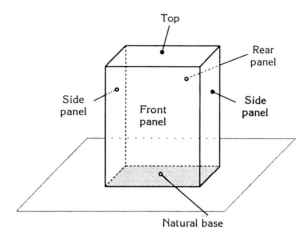

Figure 6-6 *Generic package standing on its natural base*

1. General Packages

As a general location concept, the bar code should be placed on an item's natural base, the base on which the product naturally rests on the shelf. This is recommended even though customers cannot see it, as on boxes, square-based bags, packs, etc.

Should the natural base be unavailable or unable to be printed on, the next best choice will be to print the bar code as close as possible to the natural base as in the following suggestions.

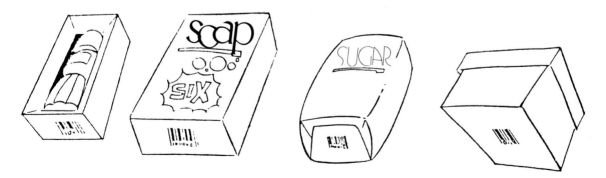

Figure 6-7 *Symbol location on the natural base*

89

2. Natural Base Unavailable

When the package is unstable, relatively shapeless, when printing might be damaged or the natural base cannot be printed on for any reason or because it is a sealing area, the second approach is to place the code on the back panel, toward the lower left if possible, like on some milk and dairy packages where the base cannot be used, or nonrigid bags (coffee, powders, grated cheese).

Figure 6-8   *Back panel, lower left corner*

3. Base and Back Panel Unavailable

One side panel will be used in this case, placing the symbol as low as possible. Examples: milk and dairy packages, fruit juice in general, brick packages.

Figure 6-9   *Side panel, lower part*

4. Base, Back and Side Panels Unavailable

This would be the only, and very unusual, situation in which the top is used. This option should be avoided whenever possible. Examples: some cosmetics, jam and items where all surfaces other than the cover do not allow printing the code.

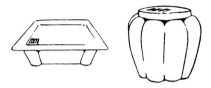

Figure 6-10 *Package top*

5. Baseless Packaging

In packages with no natural base (such as pouches, envelopes, bags), the code is located on the back panel, to the center if possible, and low when the product has a natural horizontal position. If content is liquid, code should be located at mid-height. Examples: Potato chips, coffee, grated cheese, soup or juice pouches.

solids   liquids

Figure 6-11 *Back panel, center/bottom half*

6. Flexible Packaging

In many vertical form fill seal and horizontal form fill seal flexible packages there is a seal along the back panel, and other upper and lower seals. Since modern technology allows printing in these areas, it is advisable to place the code (including quiet zones) on the back panel, away from the sealing areas, as most flexible packaging films will shrink because of heat and pressure, altering symbol dimensions.

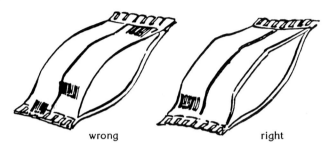

wrong   right

Figure 6-12   *Back panel, away from sealing areas*

7. Blister and Skin Packaging

The symbol should preferably be placed on the lower left corner of the back panel. If this is not possible, then the symbol should be placed on the front panel, provided that package thickness is less than 1/2″ (12.5 mm) so the distance from symbol to scanner is less than this. When the distance is more than 1/2″ the symbol should be placed on the surface of the item rather than on the front panel.

     On blister or skin packaging, the code should not be printed *on the blister or skin film* because on heat-activated thermoplastic, symbol dimensions would easily be altered. The same approach applies to polyethylene and other hot water or air shrink films and stretch films.

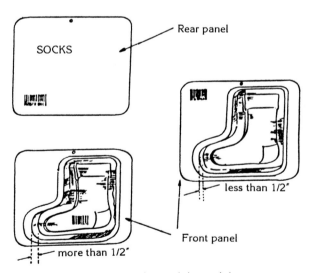

Figure 6-13   *Back panel, lower left corner*

8. Tubes

Made of aluminum, plastic or composite laminates, such as those used for toothpaste or cosmetics, tubes will bear the code printed close to the base, where the surface is flat rather than cylindrical. If the tube is always sold inside another package, the symbol need only be on the outside.

Figure 6-14 *Flat part close to edge*

9. Bottles and Cans

These items are not usually encoded on the natural base; their back part is often the most suitable, particularly if there is a label on the back where the symbol could be easily located. If there is none, the front label can be used.

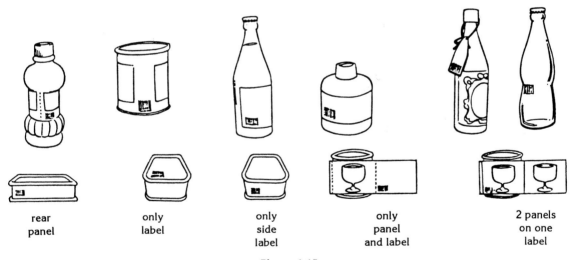

| rear panel | only label | only side label | only panel and label | 2 panels on one label |

Figure 6-15

However, some difficulties may arise:

- Front label has no room available for the code, as it must list components, numbers, and other information required by law.
- There is no label at all, and silkscreen or other printing system does not ensure minimum symbol parameters, e.g., reusable glass bottles where permanent friction, recycling and handling would end up destroying even good printing.
- There is no other label, base unavailable and top is too small.

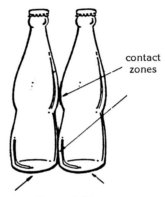

contact zones

Figure 6-16

In all the above situations, printed labels could be applied to the back panel, at the bottom. Correct label selection is critical as maximum adhesion on glass (or corresponding surface), resistance to humidity or low-temperatures will have to be ensured.

Labels also must dissolve easily when reusable bottles go through industrial washing.

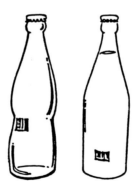

Figure 6-17

When bottles have a tag hanging from the neck, like some wines and liquors, the code can be printed on the tag, but only when no other choice is possible and provided the tag cannot be removed.

For economic reasons, whenever there is a printed label and available, space it should include the code.

The code must not be printed on a narrow label wrapped around the bottleneck.

Figure 6-18

On cans or cylindrical-shaped containers with two main panels, the code should be placed between both, on the lower section.

Figure 6-19

When the package has only one label and this is the only possible place for the symbol, it should be printed on the lower left corner, e.g., glass and plastic bottles, cleansing agents and personal care items.

Figure 6-20

On grooved and uneven surfaces, the code should not be placed across grooves. If this cannot be avoided, bar length should be extended to guarantee maximum bar exposure to the scanner.

right   wrong   right

Figure 6-21

Low, round containers and other cylindrical shaped packages, such as those used for yogurt, margarine, ice cream or juice, follow the same criteria:

a.   on the natural base
b.   on a side, lower area
c.   on the lid 1/2'' (12.5 mm) maximum depth

In truncated, cone-shaped containers, bars should be placed horizontally.

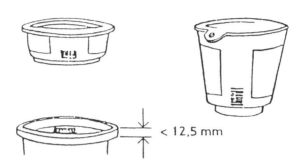

Figure 6-22

10. Labeled Products

Following the same concept, the code can be printed together with tags or labels. This system is often preferred for clothing and special sales where items are sometimes grouped and repackaged at the supermarket or retail store. The coded label is then stuck on the item before wrapping it in a standard plastic, shrink, or stretch bag. (See Labels in Chapter 9.)

Figure 6-23

11. Multiple Packaging

The code should be placed on each consumer unit according to how the item will be offered by supermarkets and distributors at their point of sale. Each unit will bear its original item code if sold individually, and the set will bear another code if items are to be sold together as a set.

Both situations can occur simultaneously, so individual codes should not be visible while items are arranged as a set. The set should be encoded on its natural base, usually a tray, frame, pack or box.

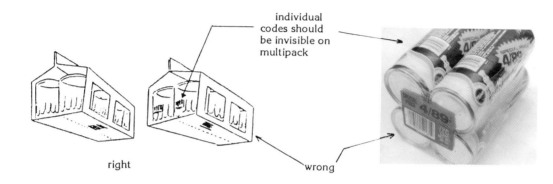

individual codes should be invisible on multipack

right   wrong

Figure 6-24

When each item is sold individually, individual codes should only be visible at the point of sale, once the original set pack is destroyed. If items are only sold as a set individual items will not be encoded.

In supermarkets, cigarette cartons containing 10 packs will only bear their set bar codes while individual packet symbols are invisible from the outside. On small items not sold individually, such as candies, the same concept will apply.

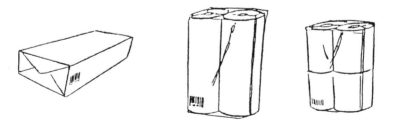

Figure 6-25

12. Tobacco and Cigarettes

The code for cigarette cartons should be printed on the back panel (where health warnings appear), horizontally centered. Large symbols like UPC-A or EAN-13 can easily be located.

If necessary, individual packets will bear their symbols on the side panel (opposite the health warning). This small packaging usually requires a reduced bar code like UPC-E or EAN-8 for marketing reasons.

Large cigar boxes are encoded on the base; small ones on the back panel. The same applies to tobacco cans and pouches.

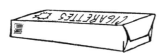

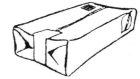

Figure 6-26

13. Fruit, Vegetable and Meat Packaging

For these and other foods priced according to the amount being selected, an electronic scale should be used capable of weighing, calculating price, printing bar code and all data on a pressure sensitive label that will also indicate unit price, item name and date.

In this case supermarkets can use their own in-store code (see Chapter 4-E), since these products will be sold at the same place they were labeled.

Figure 6-27

14. Textile Products

Clothing should bear an individual code for each size, color, presentation, quantity and model. Although the selling price could possibly be the same, shelf rotation for these items is slower and prices can change, as with seasonal sales. Besides, this system is the only way to gather reliable statistical data per item, color, size, etc.

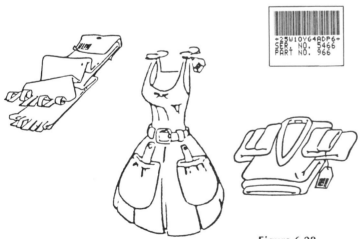 

Figure 6-28

If clothing manufacturers do not supply item codes, the supermarket can choose in-store codes as explained above, using on-site printed labels made by small, hand or desk bar code printers when necessary (see Labels in Chapter 9).

15. Continuous Packaging

In continuous packaging systems where substrate printing is not in register with the packaging machine, the symbol could be truncated at the sealing area or the package edge (A). The code should then be printed with longer than standard bars, as this will not affect scanning (B). The symbol should not be placed close to edges, but preferably located toward the center of the package.

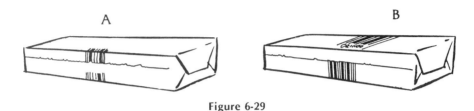

Figure 6-29

16. Cylindrical Packaging

When a natural base is not available, the code should be placed on the lower corner of the back of the item. Whenever the surface the code will be printed on is cylindrical in the final item form (the way it will go through the point of sale), the direction of the bars becomes critical. If package diameter is large enough, bar direction will make no difference because the surface will look flat to the scanner. But if the diameter is small (like a pencil) and bars are parallel to the cylinder axis, the scanner will only find a few bars and the symbol will be useless. This is why small-diameter cylinders must bear symbols with bars transverse to the cylinder axis so the scanner will see a small portion of all bars.

From a technical point of view, the following consideration should be made; when the cylindrical package is placed on a flat surface, looking from the base, the code will show a circular ring section with code center joining the flat surface and one of both tangents (first or last guard bar) forming the α angle as shown on figure 6-30.

a.   If the α angle is narrower than 30 degrees, bars should be arranged vertically (parallel to axis and perpendicular to base).

b.   If the α angle is wider than 30 degrees, bars should be arranged horizontally (perpendicular to axis and parallel to base). This choice is preferred whenever possible.

Since the item's diameter is fixed and not likely to be changed, code size and direction will be adapted to diameter d.

Maximum magnification factor to be used will depend on the α angle. In (a) it should be kept lower than 30 degrees. For each apparent diameter in package curve, a maximum MF will correspond.

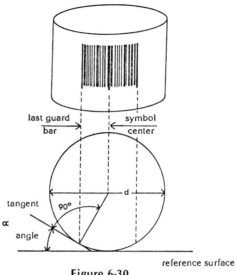

last guard bar · symbol center

tangent · 90° · α · angle

d

reference surface

Figure 6-30

- Cylindrical UPC-A & EAN-13, MF charts

The following curve allows for quick identification of maximum and minimum magnification factors suitable for each code and package diameter for UPC-A and EAN-13 systems.

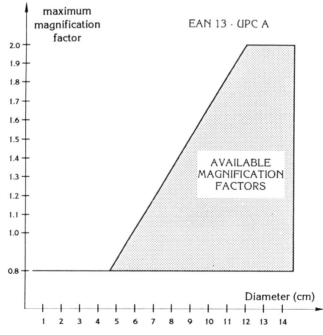

Figure 6-31 *Diameters are expressed in cm*

• Cylindrical UPC-E & EAN-8, MF charts

The following curve allows for quick identification of maximum and minimum magnification factors suitable for each code and package diameter for reduced code systems.

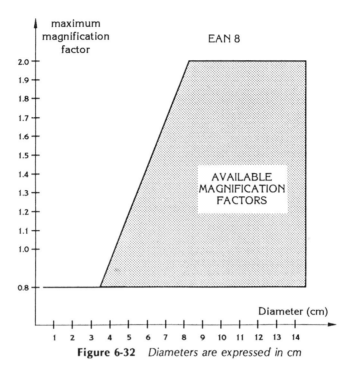

Figure 6-32 *Diameters are expressed in cm*

• Cylindrical UPC-A & EAN-13, symbol position charts

The following curves show if item belongs to group "a" (vertical bars) or "b" (horizontal bars) and help determine diameter, MF, and α angle for UPC-A and EAN-13 systems.

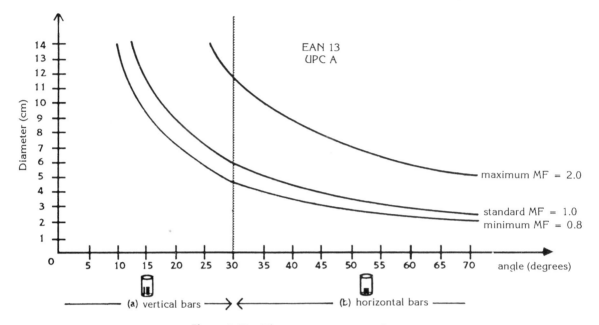

Figure 6-33 *Diameters are expressed in cm*

- Cylindrical UPC-E & EAN-8, symbol position charts

The following curves show if item belongs to group "a" (vertical bars) or "b" (horizontal bars) and help determine diameter, MF, and α angle for reduced systems.

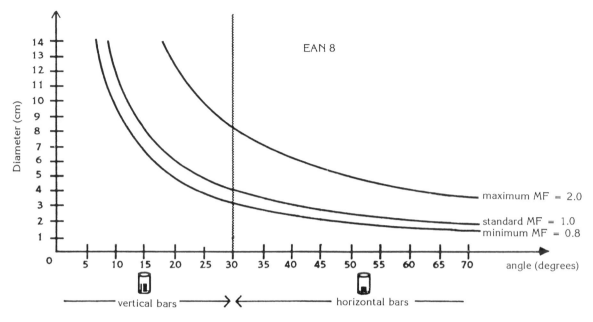

Figure 6-34 *Diameters are expressed in cm*

Obviously, if this α angle covers half the symbol, we will be considering a full-vision 60 degree angle for the whole symbol. Theory indicates that symbol position should be considered on cylindrical packaging as explained above, but this is not all. In addition, we have 0.8 < MF < 2.0 limitations, plus a minimum MF determined by the printability gauge that also establishes size limitations, plus bar direction on continuous printing systems, all of which must be considered at the same time, beginning with the package design.

Figure 6-35 shows two examples of vertical and horizontal bar arrangement in different size versions of the same item. Choice results from greater or smaller package diameter. Both UPC-A codes are correctly designed.

Figure 6-35 *Bar direction correctly designed due to cylindrical packaging*

17. Quiet Zones; examples

Minimum quiet zones specified beyond left and right guard bar patterns are a necessary part of the symbol; even though they have only spaces and no bars, an invalid scan would occur without them. Care should be taken that quiet zones are not hidden or cut, ensuring at least the minimum number of modules required which, depending on the symbol, varies from 7 to 11.

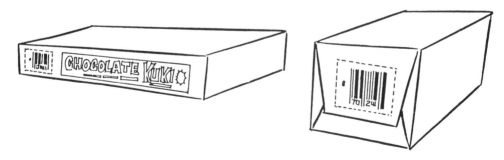

Figure 6-36 *Correct examples (sufficient quiet zones)*

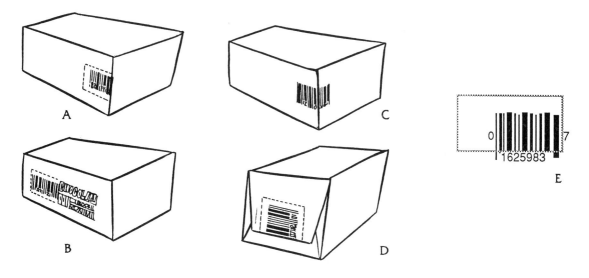

Figure 6-37 *Incorrect examples (insufficient quiet zones)*

Defects occur because of no quality control in symbol tolerances, design, printing or packaging.

A, C, D: Register errors between printing and cutting in the packaging machine
B: Register errors between text labels and code colors
E: Color register errors between bars and background

E. Code Dimension Design

1. Magnification Factor (MF)

Maximum and minimum size to be used in symbol design and printing will depend on several factors and *cannot* be chosen without taking them into consideration. They are:

    a - Symbol system specifications and magnification factor (MF) limitations
    b - Shape and features of package (cylindrical packages)
    c - Printing system and substrate to be used (printability gauge)
    d - Printing machine direction (when applicable)
    e - Space availability and economic factors

(a) "Size" is defined by the magnification factor (MF) between 0.8 and 2.0. These are the first two limitations for a UPC or EAN symbol printed on a flat package.

— A large symbol (MF = 1.4 to 2.0) is easy to print, has low tolerances, is less subject to printing errors and usually results in quick scanning, but a large symbol occupies too much space and can be undesirable for a aesthetic design reasons.

Figure 6-38 *EAN-13 (13 digits), MF =2.0, largest size permitted, most recommended*

— A medium size (MF = 0.95 to 1.4) has advantages similar to the above, takes less space and is usually preferred for marketing reasons, MF = 1.0 (or 100 percent) is the standard size and the most popular.

Figure 6-39 *EAN-13 (13 digits), MF =1.0, standard recommended size*

— A small size (MF = 0.8 to 0.95) takes very little space but is difficult to print within required specifications and tolerances, more prone to post-printing trouble and often hard to scan when not perfectly within its very strict printing margins. Acceptable quality basically obtained on thermal printed labels at this symbol size.

Figure 6-40 *EAN-13 (13 digits), MF = 0.8.*
Avoid this size when possible

— When printing space availability is critical and MF = 0.8 is not possible or not sufficient, reduced versions can be used, such as UPC-E or EAN-8. This will depend on encoding regulations, which may vary from one country to another. Once a reduced new code number has been granted, design must also follow all these requirements.

Figure 6-41 *EAN-8 (8 digits), MF = 1.0*

— Finally, when real print space available for a code is so small that all above choices fail and there is no other alternative, the code is truncated. This means that bar length is shortened. Truncation should only be applied to codes larger than MF = 1.0 and, if possible, larger than MF = 1.4. Truncation should be minimal, cutting only what is necessary. The greater the code truncation, the more difficult it is for the scanner to identify the code, resulting in a dangerous reduction of the omnidirectional scanning capability.

EAN truncation regulations are explained in Chapter 4-B; UPC does not allow truncation (see Chapter 3-C).

Figure 6-42 *EAN-13 Truncated symbol, MF = 1.6*

The following examples show actual size EAN-13 codes with magnification factors between 0.8 and 2.0. UPC-A and EAN-13 symbol sizes are the same, even though the former is 12 digits and the latter 13. Both carry the same number of bars, as the thirteenth digit is mathematically produced. (See Section 4-A-9.)

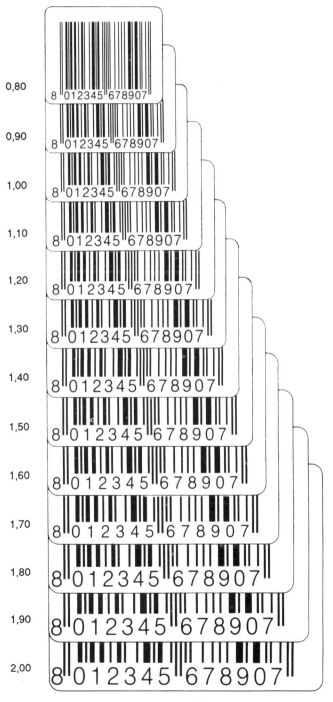

0,80
0,90
1,00
1,10
1,20
1,30
1,40
1,50
1,60
1,70
1,80
1,90
2,00

Figure 6-43

(b) The second element in limiting symbol dimensions is package shape, particularly if the surface bearing the code is not flat. Then curve diameter must be calculated in the code area (the rest of the package may be cylindrical or not). This will indicate MF values allowed and bar direction. Calculations are based on curves as shown in Section 6-D-16 (cylindrical package charts).

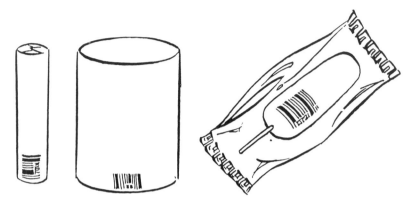

Figure 6-44

(c) The third limitation to code size is determined by the printing system and substrate. To establish these parameters, a test printing is run using the same machine, conditions and substrate, following the printability gauge tables that indicate adequate MF size according to print gain of a given printout and BWR values, as explained in Section 6-F-1. This information should be supplied by the printer for each case, package type, substrate and printing process to be used. No generalization should be made.

Figure 6-45

(d) The fourth limitation to consider is bar direction when printed on continuous systems, roll to roll such as flexography or gravure, where bars should be parallel to film edge, in the web motion direction, independent of the final package shape (see Section 6-F-3).

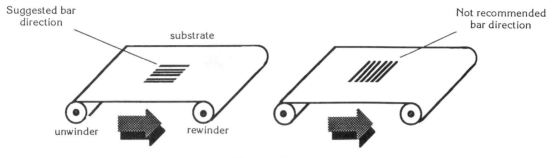

Figure 6-46

(e) Real space availability. When all previous steps indicate the impossibility of printing a UPC or EAN code on a specific package, code location should be reviewed first. If natural base was chosen but not enough room is available, we will assume package has no natural base and follow the standard procedures for code location.

For example, if the available area allows for MF = 0.90, but the system to be used is flexography on paper, the printability gauge only permits a minimum MF = 1.20. Then possible options will include relocating the symbol, changing the printing system to gravure, using higher-quality paper, and better coating. If the plates are rubber, one might attempt compressible photopolymer instead, carrying out new tests with the printability gauge until the desired minimum MF value is obtained, changing the code to a reduced version (UPC-E or EAN-8) and finally truncating the code (if EAN).

Always remember that the final purpose of a bar code is not to be designed or printed on a package but to be successfully scanned at a point of sale. To achieve this, designers and printers should repeatedly try every alternative until a satisfactory design is found within specifications.

F. Bar Code Printing Tests

1. Printability Gauge

This test indicates the print gain degree in a standard printing production, in order to determine the minimum possible magnification factor for the bar code (MF) and bar width reduction (BWR) needed to order the film master.

These are the elements involved:

Print gain is the difference between the original bar width in the film master and its final width after being printed on a certain substrate as a result of paper stock capillary, plate pressure, ink density, plate type, printing system, etc.

Print gain value allows the calculation of bar width reduction (BWR) needed on the original film master to offset average print enlarging.

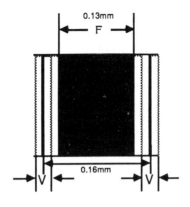

Figure 6-47

Printed bar width (P) varies according to width on film master (F), print gain (PG) and gain variation (V): P = F + PG ± V. "V" value allows for the calculation of magnification factor (MF).

Bar width on the plate (A) is calculated: A = (0.33 mm × MF) − PG, where 0.33 mm (0.013'') is the normal width of a standard module when MF = 1. For reliability reasons, bar width on the plate or printing cylinder should not be lower than 0.13 mm (0.005''), so MF values will always be such that A > 0.13 mm (0.005'').

The printability gauge is a high precision photographic film featuring two printed groups of 11 line sets each. Vertical lines are shown in one group identified A through K. Horizontal lines are shown in the other group as A' through K'. Sets are ordered according to space size between lines, from a maximum separation of 0.508 mm (0.020'') for the first A-A' pair, decreasing to 0.0254 mm (0.001'') minimum separation between lines for the last K-K' pair.

	spaces between bars (mm)
A ‖‖‖ ≡ A'	0.508
B ‖‖‖ ≡ B'	0.457
C ‖‖‖ ≡ C'	0.406
D ‖‖‖ ≡ D'	0.356
E ‖‖‖ ≡ E'	0.305
F ‖‖‖ ≡ F'	0.254
G ‖‖‖ ≡ G'	0.203
H ‖‖‖ ≡ H'	0.152
I ‖‖‖ ≡ I'	0.102
J ■ ■ J'	0.051
K ■ ■ K'	0.025

Figure 6-48 *Single Printability Gauge example*

114

As this is a precision gauge, it must always be used in its original form. It *should not* be enlarged, reduced, copied or in any way altered, as this would render it useless.

The gauge can be requested in positive or negative form, emulsion up or down, EAN and UPC compatible. Steps to follow and caution measures are the same as for film masters. The gauge must be used by the printer only, although results must be indicated to the packaging designer.

Printers, using the same printing machine and general printing conditions as in final package production, will conduct a representative sampling, printing the gauge in several positions along and across the web, on all areas where codes could later be located. The test will be more accurate if performed after running the printing machine for several hours, using the same substrate, ink and previous printing conditions with the printability gauge plate or cylinder and black ink on the last few hundred meters of whatever web was previously being printed. The gauge must follow the same preparatory steps as film master in order to obtain a reliable printability range.

Later, the printed gauge sample must be inspected with a standard magnifying glass in order to determine those line groups (A to K, and A' to K') where lines are joined for more than 50 percent of their length, identifying the first and last groups where this happens along the printed web sample, both longitudinal and transverse. These two groups of two letters are the *printability range* obtained, lengthwise (lines in the machine printing direction, or MD) and across, in the transverse direction or TD (e.g., F-G in MD, D'-E' in TD).

No contact   50% contact   100% contact

Figure 6-49 *Printed Printability Gauge bar contact as seen through a magnifying lens*

Printability range must be redefined every time any of the variables is altered. Printers will determine these values for each of the following situations:

(a).   For each printing machine
(b).   For each type of substrate to be printed
(c).   For each plate type and anilox roll
(d).   For each gravure cylinder engraving system and cell type (not entirely necessary)

In each case, printer will also consider the following variables:

(a). Printing machine conditions (temperature, pressure, tension, speed)
(b). Ink type and viscosity
(c). Pressure between plate and substrate
(d). Further printed substrate conversion processes

2. Bar Width Reduction (BWR)

Bar width reduction is done directly on the film master by the manufacturer. Film masters should be ordered by the printer specifying BWR needed for each job. This reduction (or enlargement) is aimed at offsetting print gain produced by a printing system on a particular substrate. Printability range obtained using the printability gauge (by simulating normal production on a printer) determines BWR values for the film master, according to the corresponding conversion tables.

There are two operational procedures to be performed by film master manufacturers, depending on the BWR and system required.

(a). The magnification factor is applied first, then BWR is performed (for gravure, offset, etc.), where $A - PG \leqslant 0.3$ mm (0.013″).
(b). BWR comes first, then the symbol is enlarged (for flexography), where $B - PG \geqslant 0.3$ mm (0.013″).

For this reason, film master manufacturers should know whether the printing system to be used is flexography or another system.

(a). Gravure, Letterpress and Lithography Conversion Chart

When high-absorbing cellulose substrates like paper and cardboard are printed, bars may show a slight widening as a result of ink penetration into the substrate and a capillary effect among fibers, which should be compensated for by reducing bar width in the film master.

When nonabsorbent substrates like aluminum foil are used, average bar width can be slightly lower than the original film master, as a result of surface tension and temperatures to which the substrate is exposed when bars are printed in the print motion direction. This is compensated for through bar widening on the original film master.

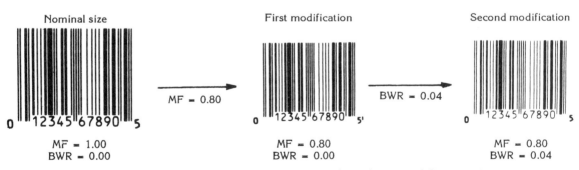

Figure 6-50 *Example of film master creation procedure when printability range is J·K*

PRINTABILITY RANGE	MAGNIFICA-TION FACTOR	BAR-WIDTH REDUCTION mm	THOUSANDS OF AN INCH
E-F	1.00	0-28 (+ 0 − 0.08)	11 (+0−3)
E-G	1.00	0.25 (+ 0.02 −0.05)	10 (+1−2)
E-H	1.20	0.23 (+0.05 −0.02)	9 (+2−1)
E-I	1.30	0.2 (±0.05)	8 (±2)
E-J	1.40	0.18 (±0.02)	7 (±1)
E-K	1.50	0.16 (+0.05 −0.02)	6.5 (+2−1)
F-G	0.90	0.23 (+0.01 −0.05)	9 (+0.5−2)
F-H	1.00	0.2 (±0.05)	8 (±2)
F-I	1.20	0.18 (+0.05 −0.02)	7 (+2−1)
F-J	1.25	0.15 (±0.02)	6 (±1)
F-K	1.30	0.14 (±0.02)	5.5 (±1)
G-H	0.90	0.18 (+0.02 −0.05)	7 (+1-2)
G-I	1.00	0.15 (±0.05)	6 (±2)
G-J	1.10	0.13 (+0.02 −0.05)	5 (+1−2)
G-K	1.20	0.11 (+0.05 −0.05)	4.5 (+2−1)
H-I	0.90	0.13 (+0.02 −0.05)	5 (+1−2)
H-J	0.95	0.10 (±0.02)	4 (±1)
H-K	1.00	0.09 (±0.02)	3.5 (±1)
I-J	0.90	0.08 (+0.02)	3 (±1)
I-K	0.90	0.06 (±0.02)	2.5 (±1)
J-K	0.80	0.04 (±0.02)	1.5 (±1)

Table 6.1. *Gravure, Letterpress and Lithography conversion chart. (EAN G.S., UPC SSM 1990)*

(b). Flexography Conversion Chart

BWR is applied first, then the symbol is enlarged to the corresponding MF value. BWR is PG/MF.

Figure 6-51 *Example of film master creation procedure when printability range is F·G*

Modern film-master-producing computerized photosetting equipment performs these operations electronically, drawing each bar to the proper width.

When rubber plates are used, the effect of an absorbing substrate (cellulose films) should be considered as well as early plate thickness reduction. For these reasons, the system usually widens printed bars very early, going out of specifications.

PRINTABILITY RANGE	BAR-WIDTH REDUCTION mm	MAGNIFICA-TION FACTOR	BAR-WIDTH REDUCTION Thousandths of an Inch
A-B	0.2 (±0.05)	2.00	8±2
A-C	0.2 (±0.05)	1.90	8±2
A-D	0.2 (±0.05)	1.85	8±2
A-E	0.2 (±0.05)	1.80	8±2
A-F	0.23 (±0.02)	1.70	9±1
A-G	0.2 (±0.02)	1.80	8±1
A-H	0.18 (±0.02)	1.80	7±1
A-I	0.15 (±0.02)	2.00	6±1
B-C	0.2 (±0.05)	1.85	8±2
B-D	0.2 (±0.05)	1.80	8±2
B-E	0.2 (±0.05)	1.70	8±2
B-F	0.2 (±0.02)	1.60	8±1
B-G	0.2 (±0.02)	1.55	8±1
B-H	0.18 (±0.02)	1.60	7±1
B-I	0.15 (±0.02)	1.80	6±1
C-D	0.2 (±0.05)	1.70	8±2
C-E	0.18 (±0.05)	1.60	7±2
C-F	0.18 (±0.05)	1.55	7±2
C-G	0.18 (±0.05)	1.45	7±2
C-H	0.18 (±0.02)	1.45	7±1
C-I	0.15 (±0.02)	1.60	6±1
D-E	0.18 (±0.05)	1.55	7±2
D-F	0.18 (±0.05)	1.45	7±2
D-G	0.18 (±0.05)	1.40	7±2
D-H	0.2 (±0.02)	1.30	8±1
D-I	0.15 (±0.02)	1.45	6±1
E-F	0.18 (±0.05)	1.40	7±2
E-G	0.18 (±0.05)	1.30	7±2
E-H	0.18 (±0.02)	1.20	7±1
E-I	0.15 (±0.02)	1.30	6±1
F-G	0.15 (±0.05)	1.20	6±2
F-H	0.15 (±0.05)	1.15	6±2
F-I	0.15 (±0.02)	1.15	6±1
G-H	0.15 (±0.05)	1.10	6±2
G-I	0.15 (±0.05)	1.00	6±2
H-I	0.15 (±0.02)	0.90	6±1

Table 6.2.   *Flexography conversion chart. (EAN G.S, UPC SSM 1990)*

The second suggested application of the printability gauge is on the production printing process, using one to three of the gauge blocks as a quality control indicator close to the bar code symbol, to check that printability range does not vary, going beyond specifications.

Positive BWR values always refer to bar width *reduction*. When BWR values are *negative*, they refer to bar width *enlargement* (very rare). This might also happen when a dark package substrate is to be used as bar color, so only the spaces or background is to be printed white; since spaces are printed (not bars), spaces could need reduction on the film master, so *positive space width reduction* might be replaced by *negative BWR*.

(c). MF/BWR Printers Chart

Once the tests with the printability gauge have been carried out, each printer can make a chart to determine minimum MF and BWR of all possible printing machines and substrates in the plant. Film master orders will then be entered by the printer, according to each job requirement, simply by referring to this chart.

ABC PRINTERS Inc.										
Printing Machine			Substrate		Print	Printability	Range		Film Master	
printing system	mach. #	anilox (lines)	type	weight (gsm)	direction	from:	to:		MF minim.	BWR (mm)
Gravure	1	150	Paper	80	L	H	I		0.90	0.13
			TiO2		T	J′	I′		0.90	0.08
Gravure	2	250	OPP	25 mic	L	I	J		0.90	0.08
					T	I′	K′		0.90	0.06
Gravure	3	150	Foil	40 mic	L	J	K		0.80	0.04
					T	I′	J′		0.90	0.08
Offset	1	150	Paper	60	L/T	G	I		1.00	0.15
Offset	2	250	Board	250	L/T	F	H		1.00	0.20
Flexo	1	150	LDPE	60 mic	L	D	H		1.30	0.20
					T	D′	E′		1.55	0.18
Flexo	2	150	Paper	80	L	D	I		1.45	0.15
					T	D′	E′		1.55	0.18
Flexo	3	250	OPP	25 mic	L	H	I		0.90	0.15
					T	G′	I′		1.00	0.15
Flexo	4	150	LDPE	40 mic	L	B	F		1.60	0.20
					T	B′	C′		1.85	0.20
Flexo	5	150	LDPE	200 mic	L	D	H		1.30	0.20
					T	D′	F′		1.45	0.18

Table 6-3 *Printers chart example*

(d). Print Gain Variation/MF chart

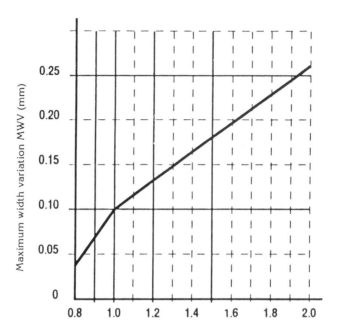

Figure 6-52   *Mininum Magnification Factor*

Maximum allowed variation (mm)	(%)	Minimum MF	Maximum allowed variation (mm)	(%)	Minimum MF
± 0.035	13.4	0.8	± 0.171		1.45
± 0.051		0.85	± 0.178	35.9	1.50
± 0.069	23.1	0.90	± 0.184		1.55
± 0.085		0.95	± 0.192	36.6	1.60
± 0.101	30.8	1	± 0.201		1.65
± 0.108		1.05	± 0.209	37.1	1.70
± 0.115	32.2	1.1	± 0.216		1.75
± 0.124		1.15	± 0.224	37.6	1.80
± 0.132	33.3	1.20	± 0.233		1.85
± 0.140		1.25	± 0.241	38.0	1.90
± 0.147	34.7	1.30	± 0.250		1.95
± 0.152		1.35	± 0.256	38.8	2
± 0.163	35.2	1.40			

Table 6.4.   *Printed bar width variation permitted for EAN & UPC symbols (NOT for film masters)*

(e). Bar Width, Frequency Charts

The following effects are very rare. They were produced in Spain some years ago, by a traditional graphic industry (H. M. Blasi SA) who dared to experiment with different bar code printing techniques, both following the EAN specifications and violating them in order to reconfirm specifications, following or disregarding them. The company tried using different printing machines on different substrates and under different film master conditions. Conclusions are very interesting and reproduced below.

In the graphs shown, values given correspond to bar width samples taken during the real printing process. Each chart represents a special circumstance, and variations on each of them serve to analyze the cause and to find appropriate solutions. Quality control was performed with an AUTO SCAN 7100 analyzer.

(f). Distorted Film Master Effects

This chart shows the effects of gravure printing on the same machine, on 110 gsm paper stock, with both a correct and a distorted (out of specifications) film master. The latter has very thin bars and deviations between -10μ and $+12\mu$ inside symbol structure. When printed, bars show a width variation of 60μ (M-P test machines) and 50μ variation (M-A 1 and M-A 2, once adjustment directions were applied).

M-A 3 frequency distribution corresponds to a different printout on machine 1, using an original film master (within specifications) and suitable for width this particular case, resulting in an acceptable of 22μ, average X = 5μ. This confirms that UPC/EAN film master specs and tolerances are correctly designed.

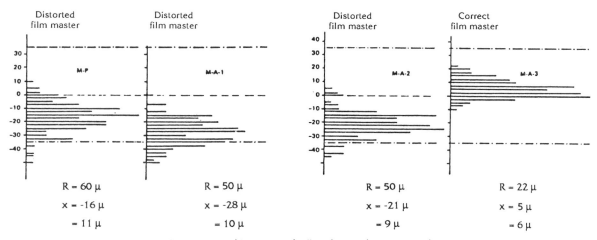

Figure 6-53 *(Courtesy of Hijos de M. Blasi SA, Spain)*

(g). Wrong MF Effects

In this experiment the wrong MF was purposely chosen, below the minimum required as per the conversion chart. The frequency distribution on the next graph represents a rotogravure printing on a 55 gsm cellophane substrate, with MF = 0.8. The tolerance of $\pm 35\mu$ was not large enough to cover the real variation of 101μ. A minimum magnification factor of 0.9 should have been selected to ensure the $\pm 69\mu$ tolerance, typical for deviations of this substrate.

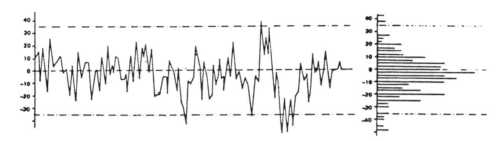

Figure 6-54   *(Courtesy of Hijos de M. Blasi SA, Spain)*

(h). Printing on Different Machines

Similar types of printing machines have different quality features influencing their print gain: Therefore the minimum MF must always be selected for each specific printing machine.

Gravure printer # 1, variation: 63 μ

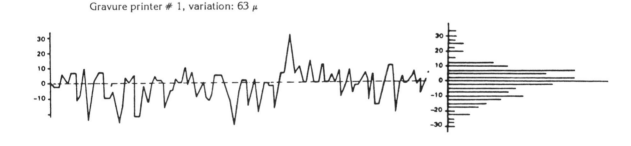

Gravure printer # 2, variation: 45 μ

Figure 6-55   *(Courtesy of Hijos de M. Blasi SA, Spain)*

122

The experiment in Figure 6-55 was performed by running the same symbol-engraved cylinder on two similar rotogravure printing machines (printer 1 and 2) on the same coated paper of 110 gsm. The quality characteristics obtained are different, confirming that printability gauge tests should be performed individually on each machine and substrate.

3. Bar Direction

When symbols are printed in continuous systems (roll to roll) such as flexography, gravure or other rotary systems, symbol bar should, when possible, be parallel to substrate edge and motion direction (MD), independent of package shape. When this criterion is not observed, TD bar widths have a tendency to become uncontrollably thicker and consequently out of specifications, rendering the code useless and unpredictably spoiling the production. When transverse bar printing cannot be avoided, higher magnification factors must be selected (MF = 1.5 − 2.0). Higher tolerances of these MF values will minimize negative transverse effects. Minimum MF and BWR to be used in every case should be tested with the printability gauge, applying the transverse printability range (e.g., J-K or H'-K'), according to print direction.

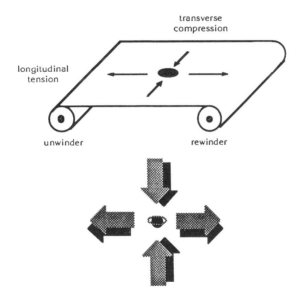

Figure 6-56   *MD tension and TD compression effects in roll to roll printing machines*

The reason for bar direction criteria in all continuous printing processes is that the different forces act on each point of a substrate going through the press. The web is constantly subject to two types of forces:

- Longitudinal tension: to maintain motion, stretching the web in the machine direction.
- Transverse compression produced on the web by the high temperature of the ovens. Shrinking is typical with thermoplastic flexible substrates, cellulose films and even aluminum foil.

A perfectly round printed dot might stretch slightly in the machine direction and compress in the transverse direction at the same time, ending up as an oval. Bars in a bar code symbol may experience this same effect, so this minimum enlargement of bar width could easily take the symbol out of specifications, whereas the same enlargement of bar length will make no difference. Bars should therefore be printed in machine motion direction whenever possible.

a. Bar Direction Effects in Gravure Printing

The following charts represent two rotogravure bar code printings on the same substrate. The same symbol was printed twice, in machine direction and simultaneously in transverse direction. The upper graph shows transverse bars, the consequence is an amplitude of 88μ and an average X of 59. The lower graph was obtained with MF = 0.9 and bar direction parallel to web edges. The width obtained decreased to 63μ and an average X of 12μ.

Even though the transverse bars might represent a minimum reduction in the amount of ink used, the uncontrollable bar direction width enlargement could also be increased by other technical problems, like the doctor blade effect in rotogravure or one-roll flexo printing systems.

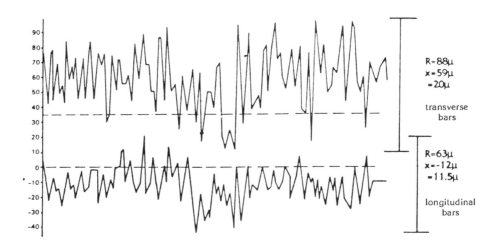

Figure 6-57 *(Courtesy of Hijos de M. Blasi SA, Spain)*

G. Color and Contrast

The scanner reads the symbol by emitting a light beam of frequency between red and infrared, scanning the symbol across the bars. Light frequency is such that it will be absorbed by bars (dark) and reflected by spaces (light). The scanner itself receives the reflected light beam, transforming reflected light into electric analogical signals and then into digital pulses as follows:

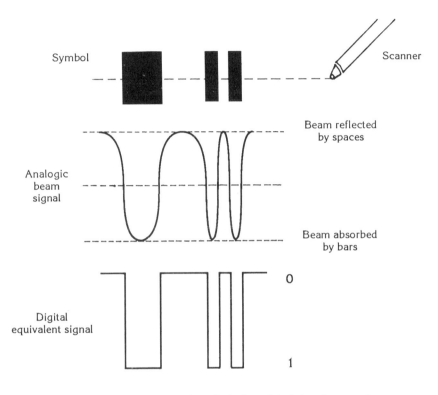

Figure 6-58 *Optical scanning and analogical to digital signal conversion*

Therefore, good performance of the symbol scanner system depends directly on the symbol's optical features (color, reflectance and contrast) as well as the scanner's wavelength (red, intermediate or infrared).

1. Reflectance

The following parameters will be defined.
- Incoming luminous flux (Frs): corresponding to photometry barium sulfate of magnesium oxide standard, where reflectance factor = 100 percent. This is used as incoming flux in the sample to be analyzed.
- Reflected luminous flux (Fr): reflected by the printed symbol sample analyzed by a precision analyzer.
- Reflectance (R): ratio between both values, R = Fr/Frs.
- Reflection density (D): D = − log(10)R.

2. Contrast (PCS)

Printing contrast or print contrast signal (PCS) is defined as the relationship between reflectance factors of light spaces Rl, and dark bars Rd.

PCS: (Rl-Rd)/Rl. (Contrast should be greater than 63 percent under EAN standards.)

3. Minimum Reflectance Difference (MRD)

MRD is the minimum difference between the lightest bar and the darkest space in a bar code. MRD = Rl(min.) − Rd(max.).
The minimum value should be: MRD > 37.5 percent when module < 0.04″ (1.02 mm).

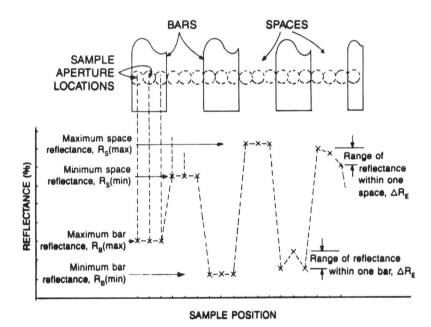

Figure 6-59   *Bar Code Reflectance chart. (AIM)*

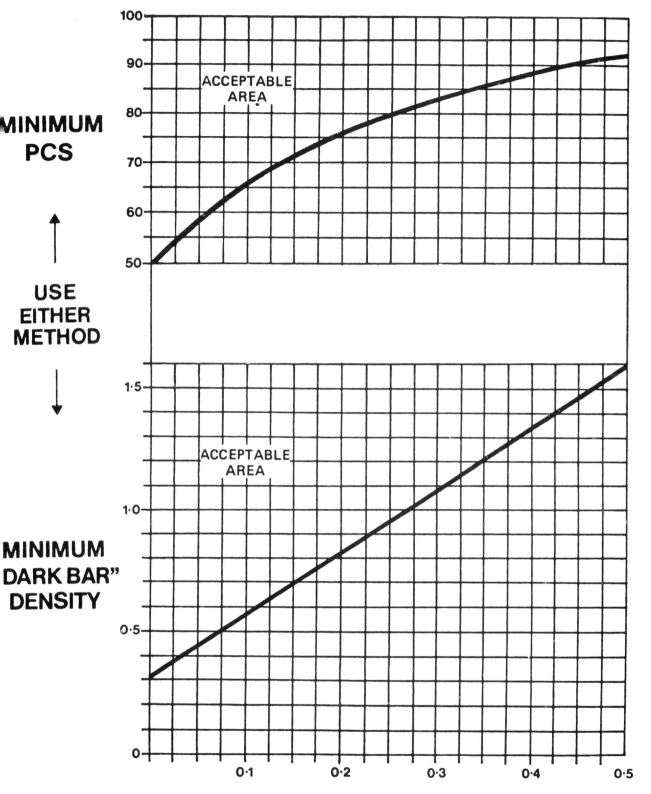

MINIMUM PCS

ACCEPTABLE AREA

USE EITHER METHOD

MINIMUM DARK BAR" DENSITY

ACCEPTABLE AREA

"LIGHT" BACKGROUND DENSITY

Figure 6-60   *Density, Reflectance and PCS chart (EAN G.S.)*

The acceptable ratio between maximum bar reflectance, for a lighter background reflectance value, responds to the following algorithm: log(10) Rd = 2.6 [log(10)Rl] − 0.3.

LIGHT BARS		DARK BARS		MINIMUM PCS
D	RL	D	RD	
0	100.0	.300	50.1	.499
.025	94.4	.365	43.1	.543
.050	89.1	.430	37.1	.583
.075	84.1	.495	32.0	.619
.100	79.4	.560	27.6	.653
.125	74.9	.625	23.7	.683
.150	70.8	.690	20.4	.712
.175	66.8	.755	17.6	.737
.200	63.1	.820	15.1	.760
.250	56.2	.950	11.2	.801
.375	53.1	1.015	9.6	.818
.300	50.1	1.080	8.3	.834
.325	47.3	1.145	7.2	.849
.350	44.7	1.210	6.2	.862
.375	42.2	1.275	5.3	.874
.400	39.9	1.340	4.6	.886
.425	37.5	1.405	3.9	.896
.450	35.5	1.470	3.4	.904
.475	33.5	1.535	2.9	.914
.500	31.6	1.600	2.5	.921

Table 6-5 *Density, Reflectance and PCS table (EAN G.S.)*

Reflectance measuring requires the following range conditions.

- The sample is illuminated using a light source with a range power distribution close to CIE light source A illuminator standard, defined by a gas-filled tungsten filament operating at a color temperature of 2856 K.
- Reflected flux photometer receptor should have a relative range sensitivity characterized by the photo multiplier in an S-4 response under specifications by Joint Electron Devices Eng. Council (United States), using a WRATTEN 26 red filter meeting nominal specifications.
- Incoming light is centered at 45 degrees from vertical for the code sample being analyzed.
- Sample opening will be a circle of 0.2 mm diameter.

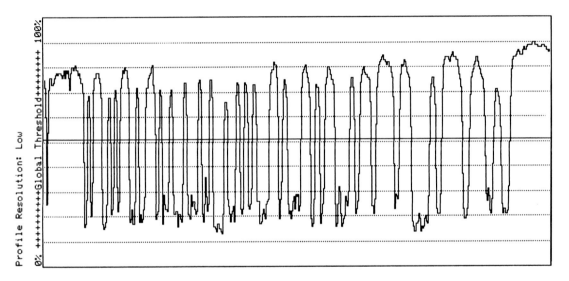

Figure 6-61 *Scan Reflectance Profile (Courtesy of RJS Inc.)*

4. Gloss and Opacity

There are no restrictions regarding symbol gloss or opacity, provided that reflectance and printing contrast values meet specifications.

5. Colors and Tones

Some colors with high red or yellow components are likely to produce a very high reflectance of red or infrared scanner light, the same as white. Certain dark colors will give the opposite effect (low reflectance or high absorption) when illuminated by a 633 nm red light. The bar colors recommended for symbol printing on most substrates are:

Black: always the most suitable
Blue: with a high cyan content
Green: with a low yellow content
Brown: dark only, with a low red content

Cyan content in a color results in the darker tone seen through a Wratten 26 filter and thus in low reflectance to red light. A high carbon content in color inks is specially suitable both for red and infrared scanners. Bar reflectance should be very low (high absorption), always Rd (max.) < 30 percent. Practical values should be: Rd = 1-18 percent.

The colors recommended for the spaces are:

White: always the most suitable
Yellow: very good
Orange: with no components from other colors
Red: with no components from other colors

Inks used for background (spaces) should be low in gloss to allow for contrast. Space reflectance should be high enough (low absorption) for symbol contrast to be higher than the minimum EAN/UPC specification of 63 percent. Minimum space reflectance should always be Rl (min.) > 31.6 percent.

6. Color Combinations You Can Use

Figure 6-62 shows possible combinations of the four recommended colors for bars, on the four suggested background colors, all of which will produce a high contrast, over the 63 percent UPC/EAN specification, suitable for symbol printing and red light scanning on most substrates.

black on white

blue on white

black on orange

blue on orange

green on white

dark brown on white

green on orange

dark brown on orange

black on yellow

blue on yellow

black on red

blue on red

green on yellow

dark brown on yellow

green on red

dark brown on red

Right and wrong color selections on these pages can be viewed through a WRATTEN 26 red filter, in the same way a scanner will look at them.

Wrong bar color selection examples

Figure 6-62

7. Color Combinations You Should Not Use

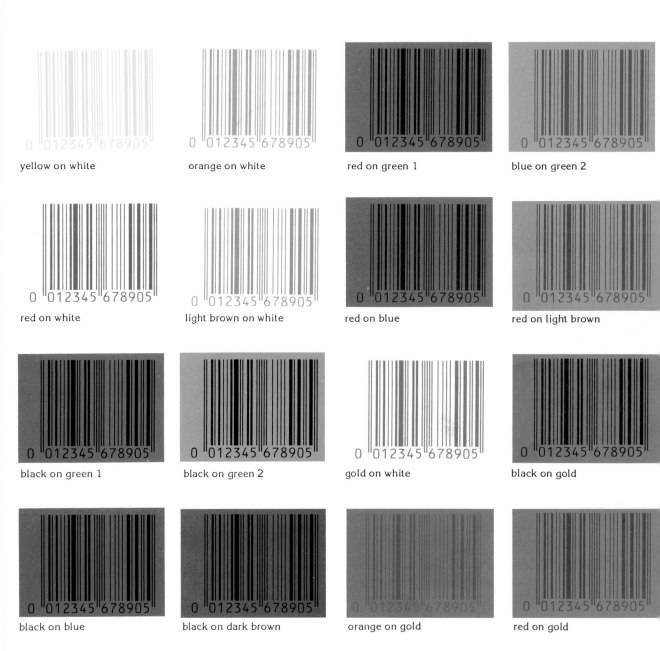

yellow on white

orange on white

red on green 1

blue on green 2

red on white

light brown on white

red on blue

red on light brown

black on green 1

black on green 2

gold on white

black on gold

black on blue

black on dark brown

orange on gold

red on gold

Figure 6-63

8. Metallized and Metallic Colors

These substrates and/or inks are *not* recommended for use in the symbol, either as bars or as background, and should be avoided whenever possible, replacing the metallic colors by those indicated above. The metallic package surface should not be used as background. A white window should be printed behind the bars as replacement. The same criteria apply to flexible packaging high-vacuum metallized films.

Metal pigments included in ink formulations, as well as metallized and metal substrates, will reflect the scanner beam with enough reflectance strength but with an unpredictable angle, so that the scanner probably will not collect the whole reflected beam.

9. Half Tone

The use of reticules should be avoided whenever possible, otherwise all half-tone printing will have to meet contrast specifications as indicated above.

10. Symbol Darkening

When reticules are used, reticule PCS should meet minimum specifications and be at least equal to the PCS of the symbol. Reticule space width should not be greater than twice its own bars.

Figure 6-64   *Reticules willl darken symbol background. (UPCN SSM. 1990)*

11. Spectral Distribution Charts

The following charts represent the different reflectance of various substrates and printed inks. The A-B line indicates the monochromatic band of 633 nm that belongs to the red scanner light spectrum, or the symbol image we can obtain through a Wratten 26 filter; the contrast is measured precisely at this wavelength.

- Figure 6-65 shows the reflectance and excellent contrast of 96 percent (A-B) of a white color printed on a transparent substrate (reflectance line 1), and blue printed on white paper (reflectance line 2).

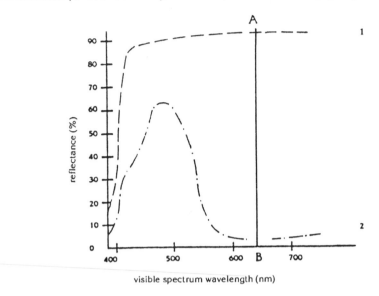

Figure 6-65 *(Courtesy of Hijos de M. Blasi SA, Spain)*

- Figure 6-66 shows the contrast reduction to 92 percent when gray paper replaces the white paper from the previous drawing. White on gray base (line 1) and blue on white base (line 2).

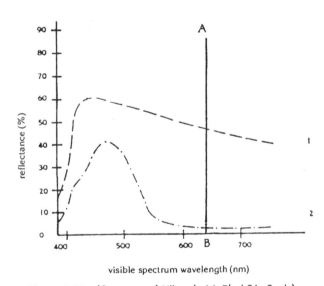

Figure 6-66 *(Courtesy of Hijos de M. Blasi SA, Spain)*

134

- Figure 6-67 shows the following data:
  - White board (line 1).
  - Red printed on white base (line 2), invalid contrast of 7 percent.
  - Gold printed on white base (line 3), invalid contrast of 47 percent.
  - Black printed on white base (line 4), excellent contrast of 93 percent.

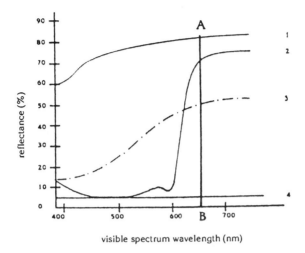

Figure 6-67   (*Courtesy of Hijos de M. Blasi SA, Spain*)

- Figure 6-68 shows the different spectrums produced by the same green color in different tones proposed for the bars, all of them on a white substrate.

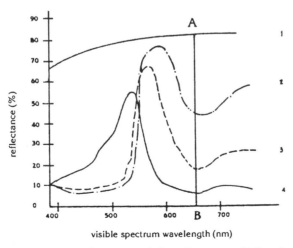

Figure 6-68   *Green color tone variations (Courtesy of Hijos de M. Blasi SA, Spain)*

- White base (line 1).
- Green (yellowed), invalid contrast of 52 percent (line 2).
- Green (medium), valid contrast of 80 percent (line 3).
- Green (blue), excellent contrast of 92 percent (line 4).

H. Printed Substrates

To be more easily understood, substrates allowed for bar code printing can be divided into three groups. This analysis is based on bar code and package area where the code is printed rather than on the package itself, and the package's capacity to contain certain products or its external properties. The following groups were defined for bar-coding purposes only:

a. Flexible Packages

Includes all packages where bar code is printed on flexible film according to the list described in detail in Chapter 7A.

b. Semirigid Packages

All packages where code has been printed on a sheet of thick plastic or cellulose film thicker than 100 micron. Packages produced by blow molding, blowing, extrusion-blowing, etc.

c. Rigid Packages

Packages where the code is printed on a rigid surface, generally glass, metal or plastic, thicker than 500 micron (see Chapter 7B).

If the code is printed directly on a rigid surface, it will be regarded as belonging to this later group. Otherwise, if printed on a label stuck to a rigid surface, it will be considered to be flexible.

1. External Transparent Code Surfaces

In all three cases the code must be designed and analyzed according to its final shape when scanned at the point of sale. This may differ from the package's shape when it was originally designed and printed.

A code printed on any substrate later coated by a transparent material (flexible film, rigid material or gloss coating) should be within specifications and meet contrast and reflectance requirements with the transparent layer on. This includes flexible packages coated with a transparent film and items packed at supermarkets in stretch film, shrink film, or in polyethylene bags. This should be considered from the design stage.

2. Transparent Packaging Behind the Bars

Transparent packages, where bars have been printed using package contents as background color, carry a high risk of going beyond contrast specifications. Bar color should be dark and correct for bars, light color and background contrast should also be analyzed once the package has been filled. A uniform background on a filled package is very unlikely due to movement of contents. Thus a non-printed background is not recommended in a transparent bag, a second suitable color should be used for background and spaces.

In some dark beverage bottles where text is printed in white, no white bars or dark background can be used, so a probable solution is to use a label or white as background and a second printing in black for bars. The same applies to transparent bags for pasta, cookies, spices, and other items when the content colors are unknown. Background should be printed in a light color to avoid transparency. Also the natural trend for transparent substrates to reduce code contrast and increase reflectance should always be considered when designing the package.

Figure 6-69   *Wrong examples; a white window should be printed behind the bars*

3. Printed Background Behind the Bars

A similar case is where the background is part of a printed art design. This should be avoided by printing a white window, for the code background. Another solution consists of cutting a window equal in size to the code, for each dark color (black/blue/green/brown) directly at the color separation.

Figure 6-70   *Wrong examples; window behind the bars must not include dark colors*

4. Humidity and Temperature Effects

Different transparent materials reflect and absorb red and infrared light in different ways and optical characteristics may vary with changing temperature and humidity. Analyses should be carried out under conditions similar to those ar the supermarket checkout counter. This includes temperature and humidity, as cellulose surfaces as paper, board and cellophane tend to absorb humidity through expansion, as with freezer products, or to lose it through shrinking (when they are subject to strong air conditioning, direct heat from lighting system lamps, or direct sunlight).

Conversely, plastic substrates tend to condense humidity on their surface. This is what usually happens with dairy products and refrigerated products in general, forming a false transparent lens effect on the code.

5. Packaging and Aging Effects

Mechanical effects on the package, and the code, as well as specific packaging conditions and aging effects should also be considered. Some items like dairy products and beverages are filled up at high temperatures, then they are cooled or frozen. The substrate under the code will experience high and low temperatures, expansion and contraction, humidity loss and gain. This can happen during the packaging process, but packaged raw materials may stay in warehouses for long periods and be exposed to weather changes. Items already packed may also be kept in stock or subjected tc changes when transported long distances (constant movement, shaking, temperature, pressure, UV direct sunlight, etc.).

All these variables and cases described are difficult to foresee, so routine checking conducted of supermarket products is suggested. The whole range of possibilities is covered when products from different origins and destinations are analyzed, to find out whether theoretical design procedures yield expected results when put into practice.

Packaging Printing

A. Flexible Packaging

1. Characteristics

Flexible packaging is generally made of one or several flexible films with individual thickness ranging from approximately 12 to over 200 microns. These can be plastic, cellulose, or metallic films, often printed in various colors; usually two or more films are bound together in a process known as "lamination," or they are coated with different products and slit together lengthwise.

All these processes are called flexible packaging conversion and are generally reel-to-reel systems, mostly rotary and continuous processes.

For bar coding, flexible packaging is the packaging part or section where the code is printed, e.g., a plastic or paper label stuck on a glass bottle or aluminum can.

Listed in Figure 7-2 are the most popular flexible films and processes, as well as a chart to check their thickness, performance and densities.

In Table 7-2 physical characteristics, dimensional stability and properties for each type of film are compared. Resistance to sunlight, oil, water and other factors bound to cause alterations to a printed substrate can be observed.

This information refers only to plain flexible films, i.e., uncoated or laminated films. For a more detailed description of lamination processes, see the section on conversion of flexible packaging, later in this chapter.

Figure 7-1 *Flexible packaging (Courtesy of Dinan SA)*

• Most popular flexible packaging films and coatings

CLL	CELLOPHANE: transparent or white, cellulose The oldest transparent cellulose film (1911)
CLAC	CELLULOSE ACETATE: transparent or colored, cellulose
NY	NYLON: transparent, extruded polyamide, plastic, thermoformable
BONY	BIAXIALLY ORIENTED NYLON: transparent, high dimensional stability, plastic
PET	POLYESTER: transparent, biaxially oriented, thermoplastic, high dimensional stability
LDPE	LOW DENSITY POLYETHYLENE: transparent, plastic
HDPE	HIGH DENSITY POLYETHYLENE: semi-transparent, plastic
LLDPE	LINEAL LOW DENSITY POLYETHYLENE: transparent, plastic
EVALDPE	EVA and LDPE Copolymer: transparent, plastic
PPC	CAST POLYPROPYLENE: transparent, non-oriented PP, plastic
OPP	BIAXIALLY ORIENTED POLYPROPYLENE: transparent or white, high dimensional stability, plastic
PST	POLYSTYRENE: transparent, semi-rigid, plastic
PVC	POLYVINYL CHLORIDE FILM: transparent or colored, plastic
PPR	PAPER: opaque, cellulose
FOIL	ALUMINUM FOIL: opaque, metal
PVDC	POLYVINYLIDENECHLORIDE: high barrier coating for films (SARAN, SERFENE)
MTZ	HIGH VACUUM METALLIZATION: metal vapor deposit on films and papers

Table 7-1

• Flexible film characteristics

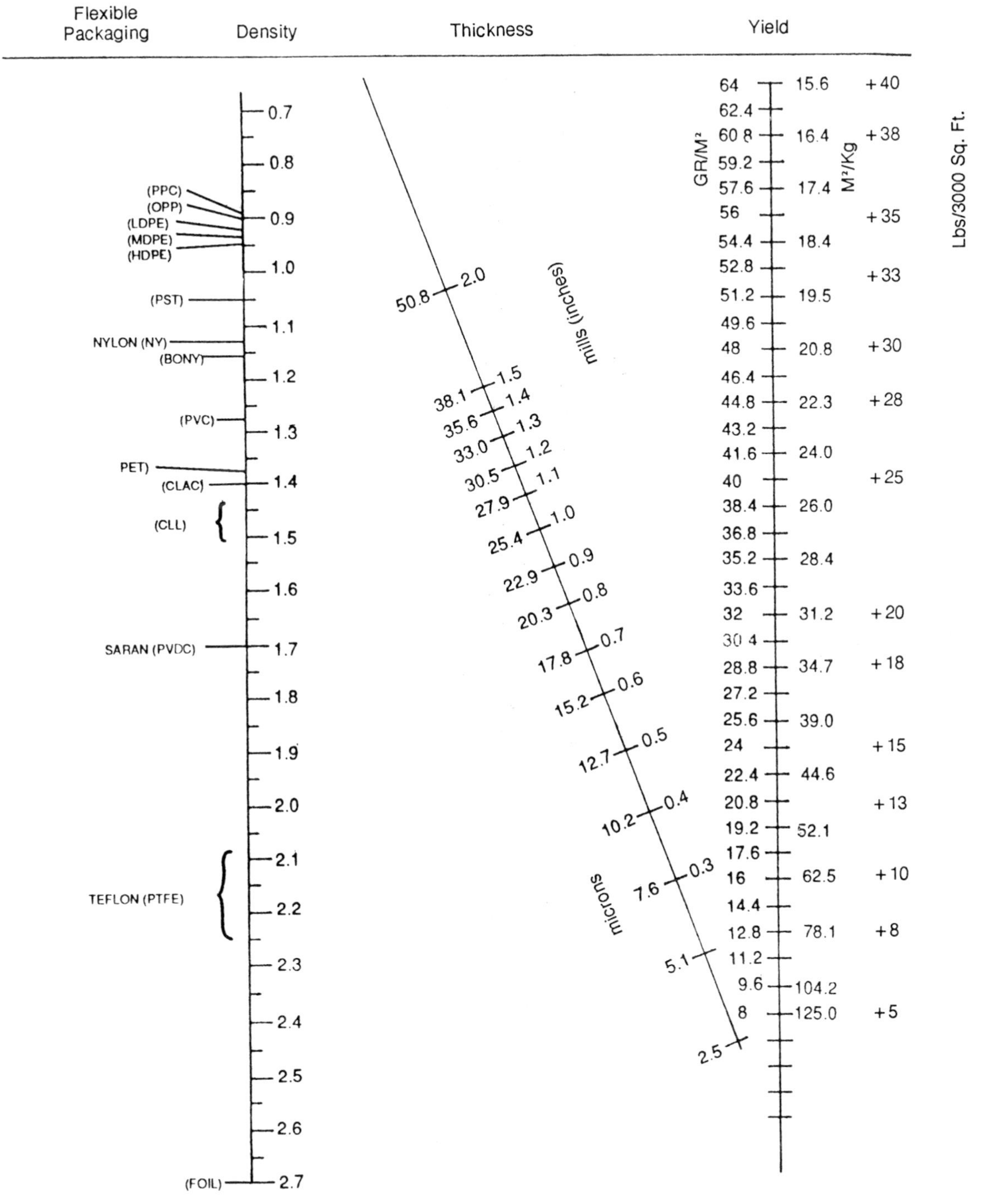

Figure 7-2   Flexible Film characteristics nomogram

- Flexible film properties compared

Characteristics		CLL	CLAC	NY	BONY	PET	LD PE	HD PE	LLD PE	EVA LDPE	PPC	OPP	PST	PVC	PPR	FOIL
																Flexible packaging film
Yield	m²/Kg	28	31	22	59	59	43	41	43	42	44	44	05	26	10	18
Weight	gr/m²	36	32	46	17	17	23	24	23	24	23	23	210	38	100	56
Standard thickness	microns	26	23	40	15	12	25	25	25	26	26	25	200	30	40	20
MD resistance	1000 x L b/ln2	18	7	9	32	20	1	2	4	1	5	8	8	2	10	1
TD resistance	1000x L b/ln2	9	16	18	36	35	4	7	7	4	10	40	12	10	10	1
Elongation, from	%	10	10	250	100	60	100	10	500	400	400	60	3	10	20	10
to:	%	50	55	550	120	165	700	650	700	800	800	100	40	500	30	30
Resistance to:																
Strong Acids		-	-	-	-	+	+	+	+	=	+	+	+	+	-	-
Strong Alkalis		-	-	=	=	-	+	+	+	=	+	+	+	+	-	-
Oil		+	+	+	=	+	-	+	+	-	+	+	=	+	-	+
Organic solvents		-	-	+	+	+	-	+	+	=	+	+	=	-	-	+
Water		=	+	=	=	+	+	+	+	+	+	+	+	+	-	+
High Humidity		=	+	=	=°	+	+	+	+	+	+	+	+	+	-	+
Sunlight		+	+	=	=	=	=	=	=	=	=	=	=	=	=	+
Water absortion 24 Hrs.	Max. %	115	8	10	9	.8	.01	.01	.01	.01	.005	.005	.1	10		
	Min. %	45	3	3	2.3	0	0	0	0	0	0	0	.04	-10		
Water-Vapor transmission, max. 100F, 90% RH	WVTR	134	40	22	11	1.3	1.5	.3	1.2	3	.7	.3	10	30		
Max. usable temperature	C	HR?	93	200	93	204	66	120	80	60	120	120	220	93	HR?	
Min. usable temperature	C	HR?	-18	-60	-60	-62	-50	-50	-50	-50	15	-50	-180	sp	HR?	
Dimensional variation with high humidity	Max. % 0	-3	-3	1.3	1.3	.5	-2	-3	<-2	1					+	
	Min. %	-.7	.2	0	0	0	0	-.7	0	0	0	0	0	0	+	0
Expected lifetime		-	+	=	=	+	=	=	=	=	=	+	+	+	-	+

Table 7-2   Compared properties of flexible packaging films

Table 7.2. data:

— Yield: (m²/Kg) area obtained per kilogram
— Film weight: (gr/m²) weight in grams of 1 m² (inverse to yield)
— Average thickness (microns): average film thickness
— Resistance (MD): Resistance to tension lengthwise, following the direction in which the film was manufactured (MD: machine direction): ASTM: D-882 standard
— Resistance (TD): Resistance to crossed tension, across the direction in which the film was manufactured (TD: transverse direction): ASTM: D-882 standard
— Stretching: percentage of stretching: ASTM: D-882 standard
— Resistance high (+), medium (=), or low (−): film resistance to internal or external physical or chemical effects, basically:
  * Strong acids: juice, preserves
  * Strong alkalis: dairy products
  * Fats, oils: chocolate, oil, butter
  * Organic solvents: present in printing and conversion processes
  * Water: depending on each case
  * High relative humidity, which might vary with the weather
  * Sunlight: infrared, ultraviolet and visible radiation
— Water absorption: percentage rate of water the film is capable of absorbing during a 24-hour period, with the resulting dimensional variation; ASTM: D-570 standard
— Water/vapor transmission (WVTR): water vapor and range that can go through a film surface under certain temperature conditions in a given time (g.mil/100 in 2/24 hs: 23-38 C): ASTM: E-96 standard.
— Operation temperature (degrees-centigrade): maximum and minimum temperatures within which the use of film is recommended
— Change in size: percent linear alteration rate at high temperature and/or relative humidity during a given time at 100 C/30 min. - ASTM:D-1204
— Estimated lifetime: Life estimation of long-stored film, printed/converted or not. The period when specifications and characteristics will not change

2. Flexography

A continuous printing system using rubber plates or modern photopolymers to transfer an image onto a flexible substrate through the use of fluid ink. Plates have a high relief engraving of the original film image to be printed. A printing unit is used for each color; this unit's plate will deposit onto the substrate, in a continuous pattern, each of the colors included in the original design.

Flexography allows for small- and medium-production continuous printing, since rubber plates, being relatively soft, warp easily and have to be replaced. The use of harder, less easily warped photopolymers considerably extends plate life.

Each basic traditional flexo printing unit consists of
   a. Ink deposit
   b. Collecting cylinder
   c. Anilox cylinder for ink transfer and regulation
   d. Plate-holding printing cylinder (metal)
   e. Pressure cylinder

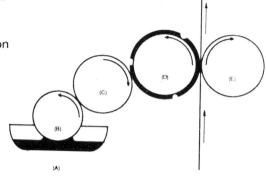

Figure 7-3 *Basic Flexo process*

Each color requires a printing unit and its corresponding hot-air drying system, including plain background, primers, heat-sealing lacquers and gloss overprint varnish. There are basically three flexography wide-web printing systems, according to unit arrangement in the machine:

• Stack-type system:
  Printing units are located on the same frame, mechanically interconnected. When more than one frame is used, register control is usually made electronically. This is the oldest system.

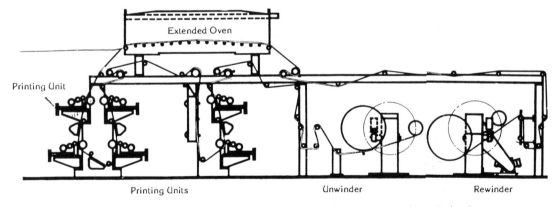

Figure 7-4 *Stack-Type 7-color flexo press (Courtesy of Schiavi-Padane)*

- In-line independent unit system:
  Each printing unit has its own frame, aligned with the next on a same surface, generally used for wall paper. This system, suitable for high-speed printing, usually has electronic color register control among units.

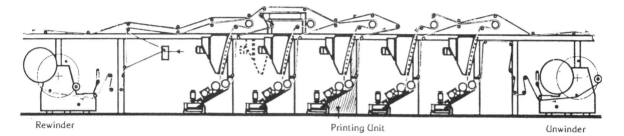

Rewinder   Printing Unit   Unwinder

Figure 7-5   *Independent Unit 5-color flexo press (Courtesy of Schiavi-Padane)*

- Central impression system (CI):
  All printing units are mounted on a single frame, and all pressure cylinders are replaced by a single central drum supporting the substrate and printing plates. This type of flexo press has been the most popular for flexible packaging for the past 30 years.

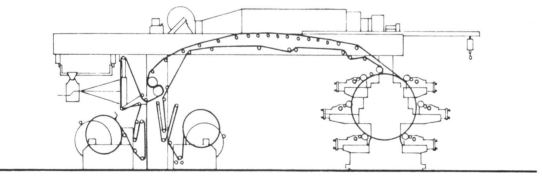

Figure 7-6   *Central Impression 6-color flexo press (Courtesy of Schiavi-Padane)*

These three flexography printing systems have different characteristics producing different results on substrates to be printed. Bar code printing demands great accuracy, and permanent observance of specifications.

The central drum system helps prevent stretching (particularly in extensible films like polyethylene), as the substrate is always around the drum. This also prevents changes in color register. The system, however, cannot prevent minor shrinking of flexible films or plate wear-out.

- Drying

In designs where two colors overlap, each color has its own fast-drying system to prevent ink from running when the next color is printed, as when symbol bars are printed on the background (spaces). Printed material finally passes through the extended drying tunnel where a heat source and air injection and extraction system evaporate all trace of ink solvent, leaving only resin and pigments on substrate surface.

Several devices help optimize flexography quality, obtaining more accuracy in bar code printing:

- Web alignment guides laterally—aligned material within requirements. Electronically operated, the guides are automatically activated by a photocell.
- Stroboscope freezing of the image during the printing process allows checking of color and register alignment. The new "image processors" will do the same task.
- Automatic tension control keeps machine direction tension, isolating printing units from wind and unwind stations, ensuring correct tension according to the film being used.
- Corona treatment is a high-voltage discharge partially breaking film surface molecules and nearby oxygen, creating a small ozone (O3) layer that oxidizes the surface, favoring ink adherence. This effect is known as "surface treatment".
- Automatic register control is an electronic system connected to each printing unit that keeps color alignment within required range, offsetting differences.
- Viscometers permanently measure ink viscosity, automatically adding solvent when necessary, to keep solid rate constant within the required density.

To complete a printing machine we should also consider an unwinder to feed raw substrate into the printing press and a rewinder to store printed material.

Figure 7-7 *New high-speed flexo press (Courtesy of Shiavi-Padane)*

- The plate

The plate is one of the most critical issues in bar code flexography printing. When engraving a soft material in high relief, it is difficult to guarantee and maintain narrow tolerance margins, as required by bar code printing in a UPC/EAN MF = 1 code. For instance, each module to be printed has a standard width of 0.013″ = 0.33 mm (1 module), according to specifications; this width must be constant through all the printing process, within ±4 mils (±0.101 mm).

Two systems are currently used to obtain flexographic plates (clichés):

a. Rubber plates: Older rubber plates are widely used. The process starts transferring the negative image on to a metal (generally zinc) plate covered by a photosensitive solution (presensitive) which, when exposed to a special light, will alter its surface to match the negative pattern. The metal plate is chemically developed removing areas not affected by light; these are later eroded with an acid, leaving the negative image in a high-relief profile. A molding press is used under controlled pressure and temperature conditions for the next two stages; the engraved metal plate is placed against another one called matrix (Bakelite); pressure and high temperature are used to transfer the image that is now formed in bas-relief onto the Bakelite. This is called the engraving pattern. The next step consists of placing the pattern in a molding press, loading it with vulcanized rubber and obtaining the final plate, which will be adjusted as it will usually have basic imperfections and uneven thickness. Since the plate is very elastic and has been engraved through several flat processes, we should bear in mind that when mounted onto the cylinder it will warp, and the image will stretch. This should be corrected prior to engraving the plate.

b. Photopolymer plates are of more recent development and have been widely accepted worldwide due to their simpler processing, greater resistance and longer life. Photopolymers are a soft, artificial monomer resin that polymerizes, hardening its surface with the action of UV light, forming long-chain molecules, which provide great resistance, flexibility and durability. These plates come in standard thickness sheets on a polyester base. Sheets are exposed to ultraviolet light that hardens the resin following negative design. Sheets are then washed, soft parts are removed, and the plate is ready for use; no adjustment is required.

There are also compressible photopolymers, where the photopolymer plate is adhered to a compressible cushion substrate on the polyester base. This provides more accuracy for fine printing jobs, as in the case of a bar code.

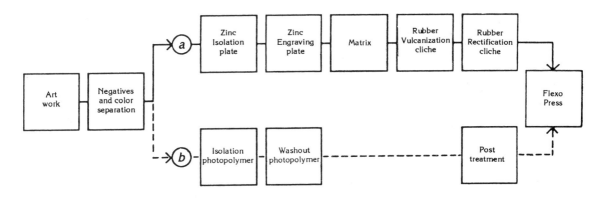

Figure 7-8 *Rubber (a) and Photopolymer (b) cliche engraving sequences comparison*

In flexography printing, care should be taken that plates, or the plate portion where bars will be printed, fall within specifications. Bars should never widen too much even if the rest of the design allows for it; bars should be kept within the alignment against the background, preventing the remaining colors and text from entering the code area.

A flexography printed bar will behave differently throughout the printing process depending on whether a rubber or photopolymer plate has been used. This is due to quick erosion suffered by rubber plates and to shape differences existing between the two.

When plates are magnified and observed from the side, we find that rubber plates show a convexity close to bar edges (necessary for Bakelite release from molding); when rubber is eroded, bars will widen out of control.

The same pattern will show concave borders if made of photopolymer (due to the action of washing brushes); it is also three to five times harder than rubber and will take longer to erode. Thus the widening effect is minimal and easily controlled.

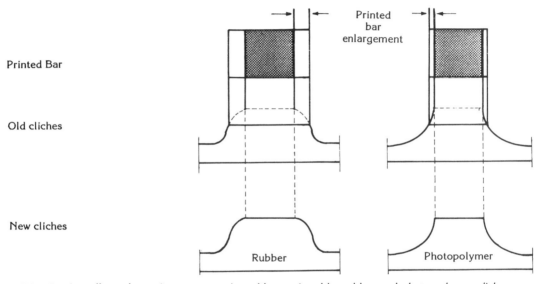

Figure 7-9 *Erosion effects shown in a cross section of bars printed by rubber and photopolymer cliches*

Whenever possible, a photopolymer plate should be used in the portion corresponding to symbol bars, and, better still, an original compressible photopolymer is recommended. The cell specification for the anilox roll is: 480 lines/inch; 5 b3mic/sq. inch; 21 mic depth for a ceramic anilox, and 360-Q-6-29 for a conventional chrome roll.

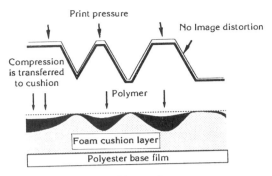

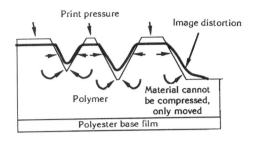

Compressible Photopolymer

Standard Photopolymer

Figure 7-10   *Cross section of compressible and standard photopolymers*

- Bearer bar

The use of bearer bars around the code and off the corner marks is recommended for bar code flexography printing. If possible, width should be greater than 8 modules. This will help to slow down plate erosion and make printing pressure on the bars uniform. The bearer bar will sacrifice its width, absorbing the printing impacts, to avoid this happening to symbol bars.

Figure 7-11   *Correct*

Figure 7-12   *Wrong upper bars section enlarged and outside specifications*

3. Gravure

Gravure is a continuous printing process using a bas-relief engraved metal cylinder to transfer ink from an image onto a flexible substrate. Gravure machines print one color per printing unit, and they usually consist of 1 to 8 units made up of:

a. Ink deposit
b. Engraved printing cylinder
c. Rubber pressure cylinder
d. Doctor blade

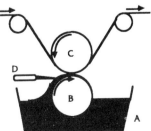

Figure 7-13   *Basic Gravure system cross-section*

149

The printing cylinder is half-submerged into ink, which enters small engraved surface cells. When rotating, the doctor blade removes ink excess leaving only the ink inside the cells, which will later be transferred to the substrate through a capillary and centrifugal effect. The system produces high quality printing, particularly on flexible material, and is especially recommended for bar code printing due to its even printing quality, higher than flexography. Obviously, this printing is more expensive and suitable for greater volumes only.

Printing units are aligned at the same level; each unit has its own hot-air drying system that evaporates fluid ink solvents, leaving only pigments and resins on the substrate surface. Due to high printing speed and continuity required, unwinder and rewinder are often automatic and can be changed without speed alterations.

Figure 7-14 shows a six-color gravure printing machine with extended ovens at first and last units; each gravure printing system is mounted on a carriage that allows for quick job exchange.

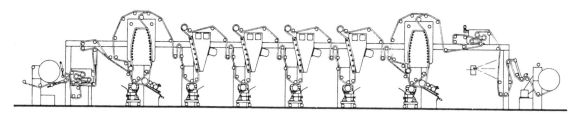

Figure 7-14 *Gravure press drawing (Courtesy of Schiavi-Padane)*

To improve and optimize performance, gravure printing machines may include a series of devices, such as:

- Web alignment guides ensure that material is laterally aligned within requirements. Electronically operated, these guides are automatically activated by a photocell.
- Stroboscope freezing of the image during the printing process allows checking of color and register alignment. The new "image processors" will do the same job.
- Automatic tension control keeps machine direction tension, isolating printing units from wind and unwind stations and ensuring correct tension according to film being used.
- Corona treatment is a high-voltage discharge partially breaking film surface molecules and the nearby oxygen, favoring ink adherence. This effect is known as "surface treatment".
- Automatic register control is an electronic system connected to each printing unit that keeps color alignment within required range, offsetting differences.
- Viscometers permanently measure ink viscosity, automatically adding solvent when necessary, to keep solid rate constant within the required density.
- Anti-static systems, where bars ionize air surrounding the substrate, and balance surface electrostatic charge.

Figure 7-15 *High-speed 8-color Gravure press (Courtesy of Schiavi-Padane)*

4. Offset

Created in the late eighteenth century, offset, or lithography, is the oldest printing system in existence. A non-continuous system, it consists of a sheet-to-sheet printing process, where the image to be reproduced comes from a non-engraved sheet; for this reason; it is also known as a planographic system.

Offset printing is based on the repellent principle between water and oil. Unlike flexography or gravure, where "image areas" are clearly distinguished from nonimage areas in height (bas-relief or high-relief), in the offset system both areas are on the same surface and are defined by the surface tension differential. Image areas to be reproduced admit oil based ink, while nonimage areas are moistened to repel oil-base ink.

Offset inks have an oil base and are non fluid. This is why a set of inking cylinders is needed to compress and spread ink on the image areas of the printing cylinder. Water is easily added, since a proper humidity balance in those areas is most desirable.

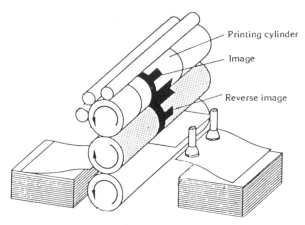

Figure 7-16

The printing cylinder supports the offset metal sheet, today's version of the ancient calyx stone, the surface of which is artificially made hydrophilous to fix the image on metal and repel ink in non-image areas.

The system gives an excellent printing quality, basically on cellulose substrates such as paper and paperboard. Bar code printing is usually trouble free when specifications are observed and printing proofs are made using the printability gauge.

In addition to the traditional offset system, several variations exist, such as dry offset that employs metal-base photopolymer plates instead of sensitive sheets. This system is used for printing bank checks, bonds and other documents on paper, as well as semirigid substrates such as PST, PP thermo-formed cups for yogurt, cheese, margarine or marmalade and for collapsible aluminum tubes containing toothpaste or similar products.

Printing proofs using the printability gauge are crucial when this last offset system is being used, to check the effect of bar code printing on different substrates.

Continuous offset printing presses (reel-to-reel, reel-to-sheet) are also available.

5. Silk Screen

This non-continuous printing system usually operates on flat surfaces (although rotary silk-screen printing machines exist, their main applications are wallpaper printing and high-quality label printing).

The image is created by a photographic process on a very well-balanced screen made of silk or any other fine fabric, where openings are left free to allow ink passage (image area), and other (non-image) areas are sealed to prevent passage. In silk-screen printing, gravure cells are replaced by fabric openings, and cell thickness is equivalent to fabric width. Both will determine the size of points and the amount of ink to be transferred onto the substrate, which may take a maximum thickness equal to that of the fabrics.

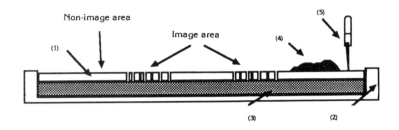

Figure 7-17 *Silk-Screen single color basic system cross-section*

The screen (1) is firmly tightened to a rigid frame directly supported by the substrate to be printed (3), flat in the example. Both are held together by a frame (2) that ensures printing alignment. High-viscosity ink (4) is applied on the screen, usually starting at one end and squeezed along the screen to the other end using a semirigid spatula (5), making ink flow through fabric openings onto the substrate.

Today, very thin steel fabrics are available; they are photomechanically engraved yielding optimum results on substrates such as rigid plastics, metal or glass. With some differences, the system is also used in the textile industry for fabric printing and high-quality labels together with UV inks.

Each color to be printed requires a silk-screen, which will be manually aligned with the other. Silk-screen printing is quite different from flexography, gravure or offset; quality and accuracy are much lower, and the amount of ink applied onto the substrate is much higher. Ink will often be thicker on the substrate; this is sometimes easily perceived by touching the ink on non-absorbent substrates.

For this reason, standard flat silk-screen is not very suitable for bar code printing. It is mainly used is in large size codes (MF => 1.6). Lower values should be previously tested.

6. Flexible Packaging Conversion Process - Lamination

Each flexible film used in packaging manufacture has characteristics that distinguish it from the rest. This also applies to the way a film reacts to printing and other processes it will undergo. Each change alters bar code parameters in a way, and we should therefore consider all processes involved in packaging up to the supermarket scanner.

So far we have analyzed different methods of printing flexible substrates, but since a universal material does not exist, different features of flexible substrates are often combined to achieve the proposed goals in terms of item protection, outside appearance and production speed. Lamination is the name given to the process of bonding two or more flexible materials together, and the resulting compound film is said to be laminated.

Flexible packaging converters are those industries capable of printing and laminating different films.

Certain items need to "breathe" through the packaging, others require gas elimination, and many items demand a complete barrier or isolation from the exterior. A barrier is the characteristic each material has to allow the passage of gases or fluids. Lamination makes appropriate combinations to provide the adequate barrier for each particular item.

Lamination systems are basically:

    a.   Duplex, dry bonding: 2 films
    b.   Duplex, wet bonding: 2 films
    c.   Triplex, wet/dry bonding: 3 films
    d.   Wax lamination
    e.   Extrusion lamination

In flexible film laminations, materials are exposed to high temperature and tension, resulting in warping (stretching/shrinkage) that alters printed code size. For this reason, we insist on code checking ar every conversion stage up to the supermarket scanner.

Lamination adhesives might be one or two component systems, mainly:

- Solvent-based polyurethane (solutions), low or high solids
- Alcohol-based polyurethane (solutions)
- Water-based acrylic, polyurethane or PVDC high-barrier (emulsions)
- Solvent-free adhesives

a. Dry Bonding

Adhesive is applied using a printing unit (B) (as if it was a color to be printed) on the whole surface of a reversed preprinted substrate like OPP (biaxially oriented polypropylene) from the unwinder (A). A hot-air injection inside the drying tunnel (C) evaporates all solvents, leaving only resins and solid matter on the substrate. At the same time, the second substrate, e.g., LDPE (low density polyethylene) is taken from the secondary unwinder (D). Both films are bonded in the nip roll (E) where the dry adhesive is activated by heat and pressure, to be later cooled in chilled rolls (F). The laminated material is collected in the rewinder (G). Laminated material obtained in this case is OPP/PE for traditional packaging use.

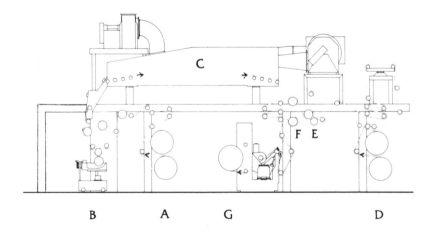

Figure 7-18 *Duplex-Dry Flexible Packaging Laminator (Courtesy of Schiavi-Padane)*

In most laminated materials, the inside is used for heat-sealing, creating the first barrier to protect contents. When three films are used, the middle film adds solidity and an additional barrier. The outside is printed and includes the bar code.

With a transparent film on the outside, the code will be seen through that film, and even through the film plus the adhesive. In the former, the scanner will take this transparency as a minor background (code spaces) darkening, slightly lowering contrast. The latter, however, has a greater darkening effect even in the case of an excellent lamination quality. If the adhesive has not been properly applied, e.g., if the anilox cylinder pattern is visible or if bonding is poor, low amounts of adhesive or low surface treatment, have produced microscopic delaminations, the results are even worse. Until the time the commercial item reaches the scanning station, defects become worse, and excellent code printings may end in an invalid scan due to defective lamination.

Whenever possible, the outside transparent material should be reverse-printed on a properly treated surface, or a primer should be used before the lamination process.

b. Wet Bonding

Wet bonding is similar to dry bonding, except that contact between both flexible materials is made before entering the drying tunnel, when adhesive is still wet in the nip roll (E). It will be evaporated later through one of the substrates, which should be a porous material, such as paper. These adhesives have a water or water-like base, and typical laminations using this system include bonding aluminum foil to paper, for butter, margarine, chocolate and cigarette box liners.

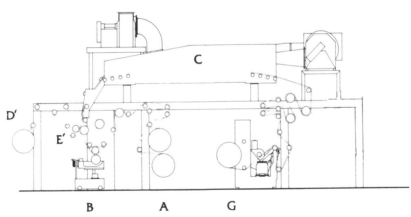

Figure 7-19 *Duplex-Wet Flexible Packaging Laminator (Courtesy of Schiavi-Padane)*

c. Dry/Wet Bonding

The dry/wet bonding machine, the most widely used, is nothing but a combination of the previons two. It gives the choice of dry lamination or wet lamination alone or both at the same time. An example would be the application of water-based adhesive on one side of aluminum foil (B), wet laminated on paper in the first nip roll (E′); solvent-base adhesive applied on the laminated aluminum foil (B′). Both are dried in the tunnel (C), and the bilaminated material is laminated with LDPE (D) in the second nip roll (E). The trilaminated material (paper/foil/pe) is then cooled and rolled on the rewinder (G).

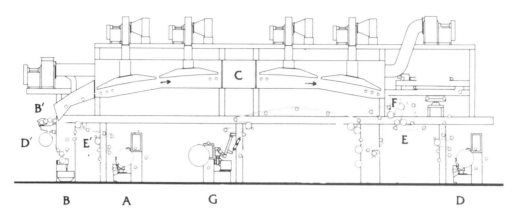

Figure 7-20 *Triplex-Dry Wet Flexible Packaging Laminator (Courtesy of Schiavi-Padane)*

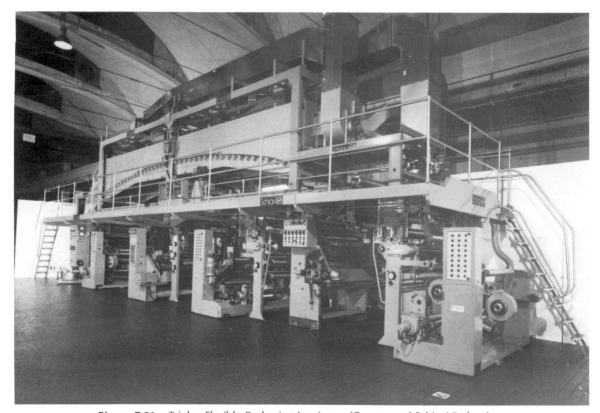

Figure 7-21 *Triplex Flexible Packaging Laminator (Courtesy of Schiavi-Padane)*

d. Wax Lamination

Wax, paraffin and resins such as copolymers of ethyl-vinyl-acetate (EVA) combine to form suitable adhesives for the lamination of certain materials; some of them are known as Hot-Melts, as they are heat-activated and do not require solvents of any kind.

Porous substrates such as paper are most appropriate for these applications, particularly when, due to wax's properties, a barrier effect against fats is desired. Laminating machines are similar to the ones described and often equipped with exchangeable trolleys for system shifts.

e. Extrusion/Lamination

Another two-flexible film laminating method without the use of solution sor emulsion adhesives consists of flat extrusion of a plastic resin, generally low-density polyethylene (LDPE), which bonds both substrates, bonding to one of them as a result of heat-blowing extrusion and pressure on the laminating nip roll. The system requires that surfaces to be laminated have adequate treatment or a primer and usually includes a heavier weight on the laminated material to add body to the packaging or to reduce aluminum foil thickness without altering the outside structure.

Figure 7-22   *Gravure press in-Line with Laminator (Courtesy of Schiavi-Padane)*

Figure 7-23 *New Solventless high-speed laminator (Courtesy of Schiavi-Padane)*

Whatever the lamination system used, the bar code should be routinely checked following lamination and again once the storing period is over. This period is required for material stabilization as it is now a flexible compound consisting of three or more elements acting together that may produce changes in size, color, contrast or reflectance in the final features of bar-code-printed symbols.

B. Rigid and Semirigid Packaging

In rigid packaging, the bar code is printed directly on a rigid substrate such as glass, aluminum cans, tin and metal containers. When there is no printed label (flexible), these containers are printed on their surface, usually using silk-screen or other indirect printing systems, with which it is difficult to obtain satisfactory bar code printing. In every case an experience using the printability gauge should be performed, then the adequate film master, magnification factor and corresponding BWR should be observed, trying to perform the printing operation with the maximum possible magnification factor.

However, rigid packaging is often used for family-sized items, that are usually heavy in their final form. For this reason, the code should be placed on the natural base or close to it, whenever possible, to prevent supermarket employees from having to manipulate a heavy item in order to obtain a successful scanner reading.

Another common problem with rigid packaging is that packages are often printed in a single color. There is little probability that the combination of color selected and the background will coincide with the contrast required.

Metal is not suitable for bars or background, nor is transparent glass. If this can be solved, code, contrast and reflectance should be carefully checked before an industrial scale printing is ordered.

Semirigid packaging is basically exposed to the situations described, with the advantage that, as these packages are mostly blown, injected, extruded or thermoformed plastics, plastic used as substrate can be sufficiently light in color and reflectance to serve as an suitable background to code bars.

Packages are generally printed using silk-screen, tampography or dry offset systems. The printability gauge test should always be performed, taking care of expansion and compression that may affect the printed code if cups are filled (heating) and then frozen (storing) or exposed to direct sunlight, as ultraviolet light may degrade color in thermoplastics, altering code contrast.

In short, each case should be closely scrutinized, and experts should be consulted. Check with your printer, test, evaluate results in a good laboratory, redefine your parameters, consult once again, check once again with your printer, test again as many times as necessary until a satisfactory solution is found. Corrugated cardboard containers, cases and boxes require special codes such as ITF.

8

Bar Code Quality Control

In bar code printing, quality control has a very special significance, since tolerances and specifications are more strict than for the rest of the printing work. An out-of-tolerance code (in dimension, color, contrast, reflectance or alignment) is not the equivalent of a similar mistake in the rest of the printing. It rather looks like a wrong text or a spelling mistake in a text. It constitutes a clear reason for rejection, with the aggravating factor that it may be rejected not just by the packing industry but by the supermarket after being packaged, which would result in losing not only the packaging but also the contents. The purpose of bar code is not to have it correctly printed but to be read by any scanner in a routine operation, at the point of sale.

Quality control methods analyzed here basically refer to routines involved in the usual design, printing, conversion and packaging transformation processes up to the scanning station in the supermarket checkout counter. All personnel involved in bar code should understand the difference between a film master quality control, a printed code quality control and scanning at the supermarket POS (which *does not control quality* of any kind; POS scanners only scan).

A. Simple Systems

The simplest way to measure code dimensions is through the use of a magnification-measuring gauge, a transparent film with a monogram indicating a specific magnification factor of a symbol and another that checks basic dimensions for a given MF. The gauge is placed on printed material and moved toward the code to be checked. Gauges provide approximate values for MF, relative dimension, sometimes also approximate module width and correlation between these data (see Figure 8-2).

Another simple system for module measuring is a magnifying lens with a base gauge graduated according to the code system and the magnification factor desired. The base is transparent and features an accurate scale and three slots corresponding exactly to minimum, medium and maximum width allowed for the module in each case. The slot is placed on the first or last bar of the symbol to check whether module width falls within specifications or not. This system is more accurate than the one described above, and only serves to measure module widths.

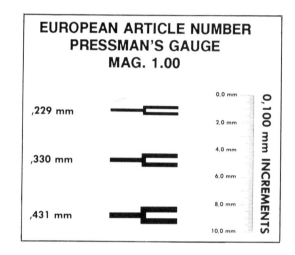

Figure 8-1 *Magnified gauge of MF = 1*

For still more accurate module measuring, a 25 to 100 X hand microscope can be used.

These systems are basically used in code and packaging design; they are also used by packaging sales people to evaluate a job and before accepting it and by marketing departments. They are simple, approximate, inexpensive manual systems, the type of system every person involved in bar code processes should carry. They cannot be used for laboratory quality control of any kind, particularly in the case of film master production.

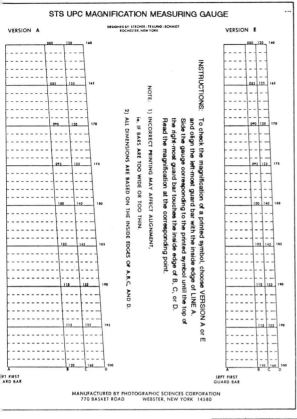

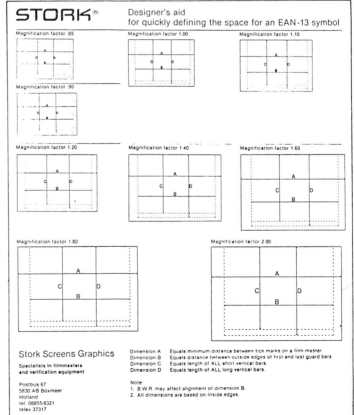

Figure 8-2 *EAN and UPC Gauges (Courtesy of Stork Graphics and Photographic Sciences Corp. Inc.)*

B. Printed Symbol Electronic Verifiers/Analyzers

Few devices are capable of strict quality control tasks that at the same time cover a wide scope of applications, are portable and easily affordable. Before evaluating them, we will discuss some of the code parameters to be checked and analyzed.

1. Symbology

Verifier should recognize Symbology as such. Symbologies usually programmed are mainly EAN-13, EAN-8, UPC-A, UPC-E (with or without add-on), code 39, interleaved 2 of 5 (ITF), code 128, Codabar, HIBC, LOGMARS, AIAG. Verifiers can often be programmed to include in-store format specifications. Older devices require external instructions about the symbol to be analyzed. More recently developed devices automatically discriminate symbology.

2. Magnification Factor (MF)

This is a very important input to make verifying equipment search its memory to find size specifications corresponding to each magnification factor or factor range against which obtained readings may be compared. Input is usually entered into the equipment by the operator; depending on code type this input is often replaced by wide-to-narrow ratio.

3. Module (M)

Module width is one of the key elements to be checked in each code, as bar and space dimension will basically depend on it. Printers are the most interested in correct printing within specifications.

The following chart indicates module width specifications for each magnification factor as well as maximum and minimum values allowed in a printed code, where permitted width error goes from ± 13.4 percent to ± 38.8 percent for EAN and UPC systems (not valid for film masters, where permitted width error is only ± 1.5 percent at MF = 1).

Magnification Factor (MF)	Module width		
	Standard width (mm)	Minimum (mm)	Maximum (mm)
0.8	0.264	0.229	0.299
0.9	0.297	0.229	0.366
1.0	0.330	0.229	0.431
1.1	0.363	0.248	0.478
1.2	0.396	0.246	0.528
1.3	0.429	0.282	0.576
1.4	0.462	0.299	0.625
1.5	0.495	0.317	0.673
1.6	0.528	0.336	0.720
1.7	0.561	0.352	0.770
1.8	0.594	0.370	0.818
1.9	0.627	0.386	0.868
2.0	0.660	0.407	0.916

Table 8-1 *All dimensions are expresed in Metric (mm), for British (INCHES) see Table 3-1*

4. Reflectance

The minimum reflectance difference (MRD) should be measured with a fixed angle formed by the scanner and the surface, as incidence and reflection angles must remain constant. A verifier often requires a black and a white set reference to compare, as well as accuracy codes printed on photographic substrates adjusted for reflectance reference. Reflectance measured by the verifier is the minimum percentage of reflected light, and main value resulting from several measurements within specifications should be given (see Figure 6-59). The minimum R value (%) = reflected flow/incidental flow.

5. Contrast (PCS)

Print contrast signal verification is basically similar to the above as regards scanner angle and surface/color reflectance needed. Contrast is the ratio between light color background reflectance, Rl (high reflectance), and dark bar reflectance, Rd (high absorption).

$$PCS = (Rl-Rd)/Rl$$

Minimum PCS values are shown in Figure 6-60. For these last optical measurements, laboratory analyzer/verifier equipment is usually required. Equipment should include red/infrared scanner, supporting base featuring the correct angle and the ability to perform sample scanning operations at as constant a speed as possible.

6. Self-check

Each piece of analyzer/verifier equipment should be delivered by manufacturers with perfectly adjusted pattern samples of different codes within specification, as well as errors, so that the instrument can be self-checked at any time. These instruments often feature a self-block system or low-battery indicator.

Bar code optical/electronic analyzers/verifiers are generally portable, battery-operated units that provide quick reading of any flat code. The units should be routinely used on already packed items as these instruments are essentially much more accurate and demanding than a supermarket scanner. QC scanners are usually pencil-type, requiring contact with the surface to be analyzed, providing basically the following four-way information:

- Quick-sound: Different sounds indicate mistakes.
- Quick-visual: Different LED.
- Visual: On a liquid crystal screen or LED, the verifiers indicate scanned code number, magnification factor and details of mistakes found or scans made, as well as additional data.
- Printed: A small printer transcribes data onto an ordinary slip of paper.

7. ANSI Guideline

The American National Standards Institute, bar code print quality guideline (ANSI, 1990) is the latest and best scientific methodology available to define procedures for measuring quality parameters of printed bar code symbols. Symbol requirements are based on the analysis of 10 *scan reflectance profiles* of each printed symbol, for a specific scanner measuring aperture and light wavelength, ranked according to academic grades, as follows:

- A (4) Superior
- B (3) Above average
- C (2) Satisfactory
- D (1) Sub-standard
- F (0) Fail

The overall grade for the symbol is the lowest obtained on any of the following eight attributes: symbol contrast (SC), modulation (MOD), defects, decodability, minimum reflectance, global threshold, minimum edge contrast (EC) and decode, the last four attributes are only subject to a pass (A) or fail (F) criteria.

C. Film Master to Printed Code Quality Control Equipment

Listed below are some of the major bar code quality control equipment suppliers and their main analyzer/verifier instruments, classified roughly as general purpose, medium or high performance equipment.

1. Stork Graphics, (Netherlands) until 1991
REA Elektronik GmbH (Germany) starting 1992

a) Master Check SC-4000-S

This is a general purpose verifier, with a red beam scanner for printed symbols only, for QC next to printing press, reception control, not suitable for film masters.

Portable, light equipment (about 0.5 kg) includes a scanner holder. Small and compact, LCD display is 16 characters + 2 LED. This is a simple, low-priced piece of equipment that scans code and automatically discriminates codes: EAN/UPC, 39, 2/5, I 2/5, Codabar, IATA tickets and 6-level bar width deviation diagnostic.

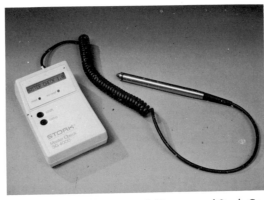

Figure 8-3   *Master Check SC-4000-S (Courtesy of Stork Graphics)*

b) Master Scan SG-4502-S

This is a medium performance, red scanner (640 nm) verifier for printed symbols, original artwork control, photographs and printings, laboratory use, printers, packaging manufacturers, supermarkets, air carriers and airports. Not suitable for film masters.

Compact, strong, desk-top type (approximate weight 1.8 kg) includes a scanner holder. Display is 64-character LCD + 3 LED. This is complex, fast, accurate equipment that is mid-priced and programmed in different languages. This unit scans and prints code, 8-level bar width deviation, print bar diagrams, static/dynamic contrast and reflectance, positive and negative and automatically discriminates these codes: EAN/UPC, 39, 2/5, I 12/5, Codabar, IATA tickets. Stores 255 measurements with 8K RAM + 8K ROM memory, external computer connection to serial port, suitable for data collection.

8-level diagnostic:

OK:	−33 to +33%
IN + (−) :	+33 to +66% (−33 to −66%)
IN ++ (−−):	+66 to +100% (−66 to −100%)
OUT+ (−) :	> +100% (< −100%)

Figure 8-4   *Master Scan SG-4502-S (Courtesy of Stork Graphics)*

Figure 8-5   *Master Scan's BAR DIAGRAM printout (Courtesy of Stork Graphics)*

c) Master Check II

This is a medium performance verifier, a new version of Master Scan in a Master Check case, released in 1991. In addition to Master Scan features it has a very small and convenient size (15 × 9 × 5 cm), is complete, light and simple. Permits 10 readings average, auto shut-off, stores data and prints to optional: Master Print II is both portable and battery operated.

Figure 8-6 *Master Check II and Master Print II (Courtesy of Stork Graphics)*

2. RJS, Inc. (United States)

a) Auto Scan 7000

This high performance, QC equipment actually "measures" the bar's width, has a constant speed and angle motor scanner head, red or infrared, available in two versions: for film masters and for printed symbols, scans from transparency and reflection, positive and negative. The Auto Scan 7000 is desktop laboratory equipment (approximate weight 12 kg) with a necessarily high price.

A complex lab system, this includes printer, bar width, analyzing and measuring each character element individually, magnification factor, density (CPI), tolerances, average deviation, contrast, reflectance, diagnosis.

The Auto Scan 7000 programs for the following codes: EAN/UPC, 2/5, I 2/5, II, Codabar, 93, 39 (AIAG, ANSI, HIBC, LOGMARS) and other options available on RJS cartridges; includes calibration standards.

This is the most popular, expensive desktop piece of equipment traditionally seen at most film master producers' laboratories, large printers and packaging manufacturers.

```
IDENT -RS                    IDENT -89
EAN #MASTER#                 EAN MASTER#
DIM IM MICROMETERS           DIM IM MICROMETERS
FLAG F1   5                  FLAG F1   8
FLAG F2   ?                  FLAG F2   4
EAN CODE  89504              EAN CODE  10773
          31979                       41102
MOD CHECK 5                  MOD CHECK 6
MAG       100                MAG       060
BWR       002                BWR      -002
BWR DIFF  002                BWR DIFF  002
LIGHT     52 %               LIGHT     52 %
DARK      85 %               DARK      08 %

   AUTO-SCAN                    AUTO-SCAN

    A   B   C   D                 A   B   C   D
 G 000 002 000               G 002 000 002
 7 002 002 005 000           4 000 000 000 000
 8 000 000 000 000           1 000 000 000 000
 9 000 000 000 000           8 000 000 000 000
 5 000 002 005 002           7-002 000 000 000
 0 002 000 000 000           7 000 000 000 000
 4 000 000 000 000           3 000 000 002 002
 G 000 000 000 000           G 000 000 000 002
 G 000                       G-005
 3 000 000 002 000           4 002 000 002 000
 1 000 000 000 000           1 000 000 005 000
 9 002 000 000 000          #1 002 000 002 000
 7 000 000 000 002           8 000 000 005 002
 9 000 000 000 000           2 002 000 000 000
 5 000 000 000 000           6 000 000 000 000
 G 000 000 000               G 000 000 000
SYMBOL IN SPEC              SYMBOL OUT OF SPEC
```

Figure 8-7 *Auto Scan 7000 printout samples (Courtesy of RJS Inc.)*

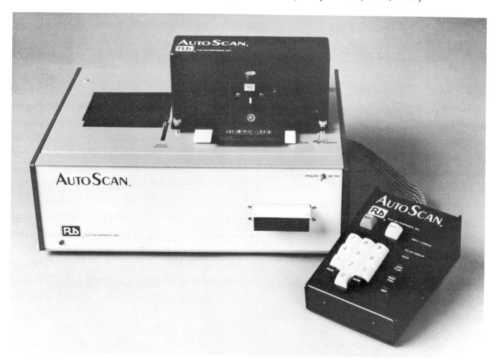

Figure 8-8 *Auto Scan 7000, with film master reading head (Courtesy of RJS Inc.)*

b) Codascan II

This is a medium-performance, desktop type, red or infrared incorporated scanner, for printed codes only, not suitable for film masters. Semiportable equipment (2,6″ × 3″ × 8″ − 19 oz), the Codascan II is medium-priced, simple and includes a printer and battery. Programming for EAN/UPC, I 2/5, 39 (AIAG, ANSI, CASECODE, HIBC, LOGMANS, 128), it scans tolerances, contrast, reflectance, average deviation. The Codascan II connects to external computers, has English or metric optional and displays 32 LCD characters, 3 LED, and has an audible tone.

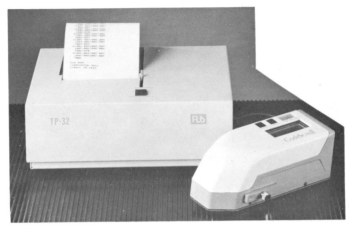

Figure 8-9   *Codascan II (Courtesy of RJS Inc.)*

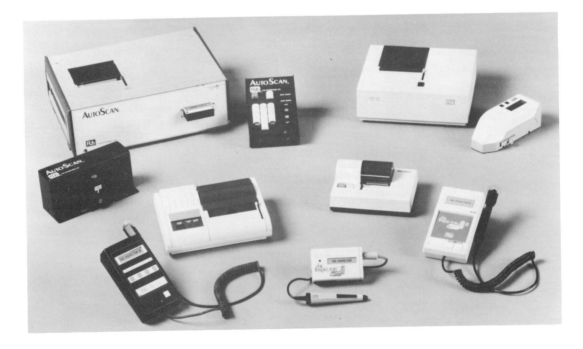

Figure 8-10   *The RJS family of bar code verifiers (Courtesy of RJS Inc.)*

c) The Inspector I, II

The Inspector is a series of four small, general purpose (models I, II) to high performance (models III, IV), portable handheld verifiers from RJS Inc. using a red or infrared wand, simple to use and low-priced. Battery-operated, the series has LCD display, LEDs, audible tone. Very light (8 to 12 ounces). Only for printed symbols, not suitable for film masters. Excellent for QC next to printing press exit as well as the lab deck or reception QC.

d) The Inspector III, IV

High performance models III and IV are new and specifically designed for the new ANSI bar code print quality verification method and AIAG, allowing for scan reflectance profile printing (graph includes: edge min. contrast, defects and decodability), aperture selection 3, 5, 10 or 20 mil as well as two wavelengths (660 nm/red and 925 nm/IR) installed in the new auto-optic scan head of model IV. Both traditional and ANSI scan profile (SRP) analysis reports are printed.

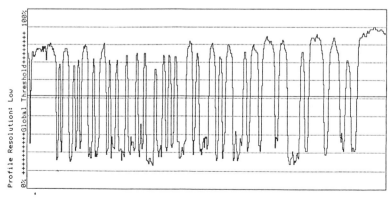

```
        INSPECTOR   III
            Revision E
     10-Scan Average Analysis

          90-115%  UPC-A
     0 45877 00585 7

   Mod Check is!............7
   Mod Check should be!.....7    PASS

        Scan Profile Analysis
   Reference Decode...............A
   Decodability..............70%...A
   Symbol Contrast...........79%...A
   Refl(MIN)/Refl(MAX).......13%...A
   Edge Contrast(MIN).......50%...A
   MODulation................63%...B
   Defects...................32%...F
   Application Compliance.........A

         OVERALL SYMBOL GRADE
     C/06/660      2.0/06/660

       Traditional Analysis

          Acceptable
        -100% Tol. +100%
        ----RRARR+++++++

   Print Contrast Signal....86% PASS
   Required PCS..............56%
   Element Refl.(MAX).......89% PASS
   Reflectance(MIN).........12% PASS

         Pass/Fail Analysis
   Passing Grade Selected.........C
   Final Results............*PASS*
```

Figure 8-11 *The inspector III, ANSI report and SRP (Courtesy of RJS, Inc)*   **171**

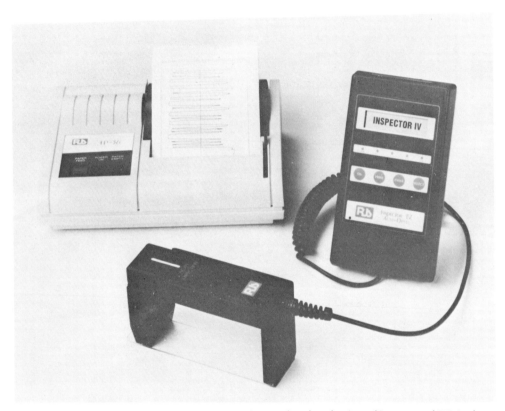

Figure 8-12 *The inspector IV, new auto-optic scan head and printer (Courtesy of RJS Inc.)*

3. Photographic Sciences Corp. (United States)

a) Quick Check 300, 400

These are small, portable, medium performance units, for printed symbols EAN/UPC, 39, I2/5, not suitable for film masters. Using a red or infrared wand type scanner, it checks MRD, contrast, reflectance and bar width specs. Light (0.8 Kg), low priced, this unit has two lines of LCD 32 character display and 5 LEDs, external optional printer is available.

More features like extra symbologies and auto-discrimination, 10 scan average, data storage and output connector to main computer are available in model 400.

b) Quick Check 500

This is the latest high-performance version of 300, 400 and older III, IV and V models. Model 500 also allows for printing and analysis of scan profile (which is the analogic scanner signal) and ANSI grade parameters, plus 16 standard industrial symbologies, 10 scan grade average. Direct module measuring, is programmable and has an optional constant speed, hands-free motorized scanner holder (Autowand) is used for flat symbols.

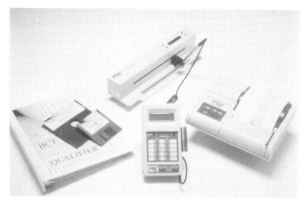

Figure 8-13 *Quick Check 500, motorized scanner and printer*
(Manufactured by PSC-Photographic Sciences Corp.)

c) Quick Check 200

This is the latest compact all-in-one release of this verifier's family from PSC. Small (7.2″ × 1.7″ × 1.2″), light (5.5 oz.) and high-priced. Offers ANSI grading with 4 apertures and 2 spectral responses plus all traditional parameters. Includes a rechargeable battery, tests main popular generic symbologies and a choice of 2 industry applications.

4. Symbol Technologies, Inc. (United States)

a) Lasercheck LC-2811

This was probably the first high-performance QC lab equipment to include a real HeNe (gas) red laser fixed scanner head. Desktop lab unit is big, heavy and expensive but very accurate as both symbol and scanner remain fixed as laser beam scans. For printed symbols only (not for film masters), the unit includes printer and external computer output. Other features include symbols' EAN/UPC, I2/5, 39 and Codabar, symbol size and module measuring, reflectance, contrast and percentage decoding.

b) Lasercheck II

Following the older desktop model LC-2811, this is a portable, lightweight high-performance unit (6.5/18 oz), battery-operated and fully automatic. Features include symbol decoding, Symbol's Scanability Trend Index (STI), average bar width (BWG), MRD, PCS, Ratio, Margins and auto MF/type selection. Only for printed symbols, not suitable for film masters.

Terminal has 128, 256 KRAM, 32/64 KEPROM, display is LCD 8 line - 20 characters, high resolution. Scan head is laser solid state red (675 nm) and requires a constant distance of 25-inch maximum from symbol surface.

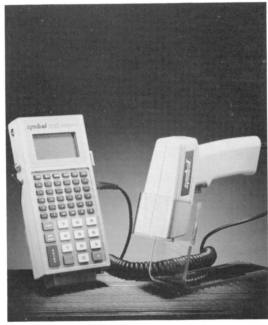

Figure 8-14   *Lasercheck II (Courtesy of Symbol Technologies Inc.)*

5. Ergi GmbH. (Germany)

a) Ergicheck SL-500, SL-600

These are portable, general purpose printed bar code verifiers, not suitable for film masters. Compact, very light (about 600 Gr.), battery-operated, they have a 16-digit LCD display and red scanner on model SL-500, with 4 LCD lines display plus red/infrared scanner on model SL-600. Standard verification routines and symbologies plus optional codes are available as well as external optional printer Ergiprint TP60.

b) Ergilaser LC-3000

This is a desktop, high-performance lab verifier, one of the few offering a real HeNe red laser motorized scan head. The truly unusual feature is the small incorporated TV monitor screen where data is clearly displayed. A small printer is also included.

This is a lab desktop instrument, not portable; it is heavy (15 Kg), maintenance-free, expensive and highly accurate, suitable for both printed symbols and film masters according to scan head type.

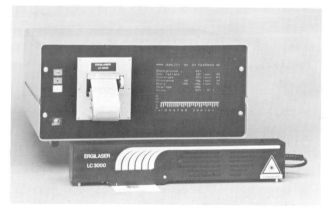

Figure 8-15   *Ergilaser LC3000 (Courtesy of Ergi GmbH)*

6. CC1, Inc. (United States)

a) Image Processor IP-100

Image Processor is a quality control concept completely different from those analyzed above. This system requires fixed installation in the printing press, being an in-line QC system for bar code's verification while they are printed. There are several available models: IP-60, IP-100c, IP100B and image Scan II.

This real-time verification system from CCI is designed for all web printing processes such as flexography, rotogravure, offset and letterpress. The system generates images of the symbol in the running web during real printing process and is completely insensitive to press speed up to 2000 feet/min. (700 m/min.) or more, permitting real in-line verification, error correction long before waste occurs and requires no special bar code knowledge from press operators.

The two main components are:

1. A master control console (Figure 8-16) which includes:

*   19" high-resolution graphics TV monitor, supplying a flicker free color image for press operator viewing (Figure 8-17).
*   Easy-to-use touch screen control (in any language).
*   Special computer.

2. A traverse installed in the press, including a color camera and strobe assembly, insensitive to ambient light. The camera traverses back and forth across the web at up to 12 inches/second (300 m/sec), measuring printed bar and space widths to ±0.001" (±0.025 mm) and comparing with UPC/EAN standards.

Figure 8-16   *Master Control Console (Courtesy of CCI, Inc)*

As the press runs at full speed, bar code images taken are displayed on the monitor with magnifications of up to 25 times. Should a symbol go below tolerance, a flashing beacon will alert a press operator, and the error will be highlighted on the screen.

Figure 8-17   *UPC-E symbol on the screen, being verified while printed (Courtesy of CCI, Inc)*

This high performance system is very expensive and designed only for inspecting printed symbols, in the printing process. Several models are available and very specialized options are software-based. A computer printout as well as other features such as color and register inspection and monitoring are the other reasons why the high-volume printer should seriously consider this important investment.

7. Coras US Corp. (United States)

a) VERISYM (r) Software

VERISYM is a PC-compatible symbol verification software system, available in different languages, for analysis, verification and statistics of printed UPC/EAN bar codes on consumer units, mainly at point of sale.

Requiring a standard AT type computer and QC high-performance verification equipment, Verisym will permit UPC & EAN specifications verification as well as manufacturer, item and printer status evaluation. Report output is on screen, printer or disk (ASCII), as an individual verification form, group list or statistics. Mailing system and automatic letter editing are supplied directly from the enclosed manufacturers item and printers data bases.

The system is designed for EAN/UPC national organizations as well as large supermarkets, retail store organizations, or service suppliers, and is currently used by EAN agencies in some countries. VERISYM(r) is a registered trademark and copyright © 1991, by the author of this book.

D. Printing, Conversion, Stock, Packaging and Traffic

One of the main functions of the analyzer/verifier is to provide printers with all the necessary information while the industrial process is going on, indicating the state of bar width and the trend it shows toward increasing or reducing its dimension. Corrective action can thus be taken before defects move outside specifications.

The equipment should indicate in a quickly and simple, way whether the code falls within specifications, whether bars are becoming thicker (or thinner), or whether printing has already gone outside specifications, and the reason for this. At the same time equipment will indicate any extra errors, check character, parity error and even contrast and reflectance if the code is not to be coated or laminated later.

Routine checking suggested while the printing process is going on, for instance when register and color tone quality controls are performed. The slip of paper provided by the analyzer/verifier scanner should be kept with the routine check form.

All printed material to be laminated should be checked once the storing period is over and before any other process. Since a transparent film, and probably an adhesive, has been added, reflectance and contrast especially should be checked, in addition to size specifications, after lamination.

Long storage periods for finished material and exposure to direct sunlight and outdoor agents may also affect material and code dimensions, particularly in the case of cellulose or low-density polyethylene substrates with no additional base. For this reason, manufacturing industries should perform careful quality controls on printed packaging material reception using analyzer/verifier equipment to be certain about optimum code conditions at this stage.

Statistical quality controls should be performed after packaging and while the distribution process is going on. Laboratory equipment will indicate successful scanning rates and average rates. Supermarket scanners do not serve this purpose, as they are not quality control laboratory instruments but a commercial tool. For instance, if a supermarket scanner performs 1 correct scan out of 1000 we may think the code is correctly printed. This would be a mistake, since a 30 percent correct scan minimum is required for a code to be acceptable, and the reading obtained gives us *no* information about the symbol quality but about the scanner's ability to read it.

This is why we insist that a supermarket scanner, or any other data collection scanner, should *not* be used for quality control testing, as it is not a quality control device.

- Calibration

All bar code analyzer/verifier instruments are usually electronic and may suffer calibration deviations for a number of reasons. The instruments should be periodically checked using suitable calibration patterns supplied with instruments.

Figure 8-18   *Wratten 26 Red filter, Contrast Check SG-1100 (Courtesy of Stork Graphics)*

• The following scan reflectance profiles show the comparison between a very good symbol and a hard-to-believe one, where bar defects are so severe that many of them look like false spaces to the scanner, causing a no-read situation.

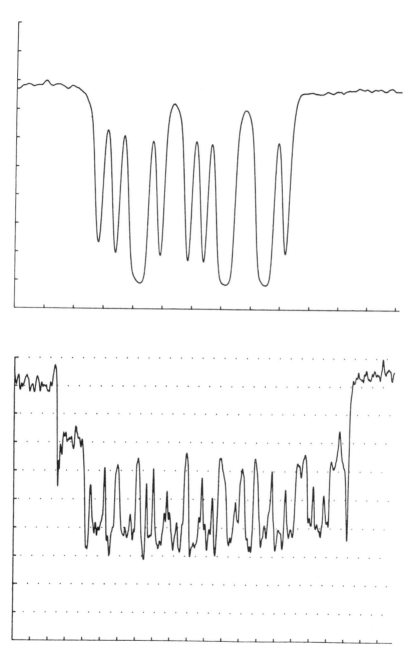

Figure 8-19 *Scan Reflectance Profiles of a Zero Defect symbol (upper) and another with Extreme Defects (lower)*

9

Bar Coded Labels

A. Bar Code Printing on Labels

For a variety of different reasons, bar codes often cannot be printed directly on the substrate or packaging; in these cases symbols should be printed on labels (whether pressure sensitive or not), using completely different printing methods and techniques from those discussed in Chapter 7.

The most common reasons to use these systems are:

- The need to re-encode wrongly coded packaging.
- The need to encode small- or large-volume productions with no code on the package or no printed package.
- The need to encode different items individually using the same stock, box type, or different presentations of the same item.
- The need for supermarkets to encode uncoded items, grouped items for a special sale, clothing of different size or color, textiles in general.
- In supermarkets, codes including weight and price of meat, vegetables, fruit, or similar items.
- The need to transcribe small-volume symbol data, with or without automatic item numbering, for in-store use, traffic or consumption, or when code changes are required.
- The need to print different code types according to item use or destination, particularly in the case of industry inputs and semi-manufactured material, under industry encoding specifications.
- Limited number of very small high-quality codes.
- Label manufacturers who offer customers printed, pre-printed and coded labels.

Basically, eight label printing systems are suitable for bar codes with some labels being pressure sensitive. Systems differ from one another in initial investment, printing costs, speed, quality, computer compatibility, paper and label systems. Before deciding on the system to use, all alternatives should be carefully analyzed.

An organization choosing one of these systems to encode its products must always keep in mind that code design standards also apply to labels, such as code location on the package (Section 6-D) color and contrast (Section 6-G) and quality control (Chapter 8).

Labels may represent approximately 60 percent of bar code substrates going through the supermarkets' points of sale.

1. Label Printing Systems

a) Thermal printing

Thermal printing goes as far back as 1919, but it was much advanced in the 1950's when the inorganic "thermofax" process was created. Present systems are either organic or inorganic and were developed in the 1960s and 1970s. Nowadays thermal paper is used everywhere in the popular FAX machines.

Thermal printing makes use of the fact that some chemical compounds, when applied to a paper surface, change color as they are heated. Selectively heating and cooling the paper surface forms the desired image. The system's core is a printing head converting the electrical pulses it receives into thermal energy, with 7 to 12 elements moving from side to side along the paper slip as it is tracked on in a mechanical movement.

This method's popularity is due to its high-quality bar code printing, format and production choice, low initial cost, computer compatibility, low maintenance cost and excellent image quality. Also available today is two-color thermal printing through temperature difference.

To ensure correct supermarket scanning, care should be taken in thermal paper selection for use in automatic label dispensers, as some thermal papers are only suitable for visible red scanners and others for infrared or invisible scanning, although some thermal papers will provide great contrast in both cases. Additional substrate tests should be carried our every time new supplies are provided.

Normal thermal paper has an average life of two years. We recommend requiring suppliers to confirm manufacturing date of every batch before accepting it. In addition, ultraviolet light may degrade thermal paper surface coating. Ultraviolet radiation is present in direct sunlight and in germicidal tubes used in critical environments such as pharmaceutical laboratories. Therefore thermal paper labels should not be exposed to outdoor agents, direct sunlight or UV tubes. Sunlight passing through a window glass represents no hazard as glass filters ultraviolet radiation. Exposure to high temperatures will result in material becoming darker and should also be avoided.

Modern thermal printers also provide thermal transfer printing (item b) in the same equipment.

- Thermal paper's main characteristics

Estimated lifetime if below 140 °F (60 °C)	Standard INORGANIC paper	Standard ORGANIC paper
Direct outside sunlight	4 weeks	4 weeks
Inside room temperature	5 years	2 years
Inside cold warehouse	10 years	5 years
Visible Red scanner compatible	most types	yes
Infrared scanner compatible	yes	not always

Table 9-1

b) Thermal Transfer Printing

This high-quality system is basically similar to thermal printing, using the same type of printing head plus a black plastic ribbon that transfers the image onto the substrate activated by heat and pressure. The principle, known as hot stamping, works with disposable ribbons. As top quality transferred images are obtained, a large range of substrates can be printed, including ordinary paper labels. Printing ranges are usually limited to ribbon width, and the latter is a major factor when costs are considered. An interesting fact is that ribbons can sometimes be removed, so the equipment can work directly on thermal paper (item a).

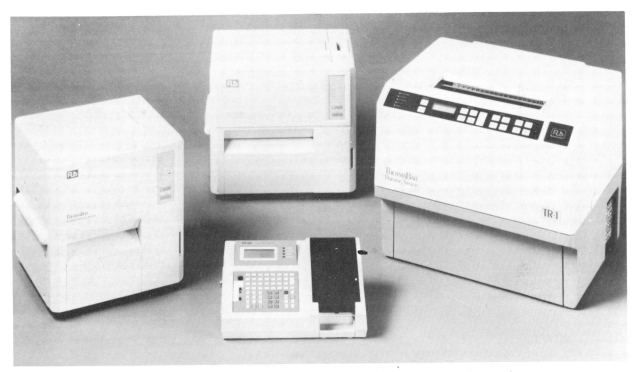

Figure 9-1   *Thermal Paper and Thermal Transfer Printers (Courtesy of RJS Inc.)*

c) Impact Printing

Impact printers usually have a turning drum on whick code elements and characters are engraved. The drum turns until the desired character is positioned following directions from a computer program. A hammer mechanism will press on the selected character or symbol, striking the paper through an ink ribbon, thus transferring the image onto the substrate. This system was popular with typewriters and telex machines. Ribbons usually have a polyester base and are disposable rather than re-inkable.

Regular paper and some plastic films can be used with this system. The printing drum is replaced to change fonts, characters or code. The equipment is quite noisy when working at high speed; it is no longer very popular.

d) Laser Label Printers

The term *laser* defines a special type of artificial, electromagnetic radiation propagation, consisting of a perfectly phased wave pack where very high energy concentrates on a light beam. This property applies to high-quality, high-definition image transfer by heat and/or pressure. A laser beam transfers the image to the printing drum which in turn transfers it to the label substrate. A corona treatment is often used to act on image transfer by way of pressure and temperature. Heat limits the type of substrate to be used, particularly in the case of compound lamination or pressure sensitive labels where heat can soften or degrade the adhesive. In the initial stages, substrate tests should be conducted for each new substrate type or batch. New technologies like ion deposition makes continuous bar code printing possible using this system, at high speed and roll to roll. Definition is very high: 300 DPI in commercial equipment and nearly 3000 DPI in special industrial equipment.

e) Ink Jet

This modern printing system can print a substrate at a distance, that is, without touching it, and works because certain ink drop types can be ejected at a distance against a target (the substrate). The process, which can be continuous, intermittent or pulse-based, is electronically controlled.

Widely used in item marking on production lines, the process can be adapted to some bar code systems, since characters and symbols do not actually exist but are stored in the computer memory. Most ink-jet systems will print only low-density symbols, which are not suitable for UPC/EAN low MF codes.

Ink jet does not transfer heat, impact or pressure onto the substrate and can even encode nonflat surfaces like corrugated cardboard, glass, metals and plastics.

This equipment can reach high printing speeds, but its use normally involves high initial costs. Low-price units are capable of printing labels, obviously at a lower speed. As it prints images dot by dot on moving surfaces, it is suitable for printing boxes and containers in-line, with large-size symbols only, usually ITF.

f) Dot Matrix Impact Printers

Basically, there are two dot matrix systems: line printers and character printers. The operating principle for both is to transfer an image dot by dot through very small hammers striking an ink ribbon against a paper label. Some bar code symbols and characters are formed through single or multiple lines of printed dots. Logically, the lowest printable module will depend on the lowest dot diameter the machine can print. Figure 9-2 shows how bar code images are formed when printed with a dot matrix printer.

The system is the most familiar in commercial and home computers; quality is acceptable for low density or large symbols if high-quality ribbons are used. Standard definition is low, about 70 DPI, though LQ (letter quality) versions provide very high definition (about 150-300 DPI) at low speed.

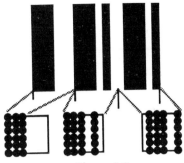

Figure 9-2

g) Electrostatic System

This system also works at a distance and is similar in concept to that used by most commercial copiers. Image generation and printing on paper has three basic stages:

1. Image is generated by transferring electrostatic charges from an ion generator onto a dielectric substrate. Charges pass through a dot matrix adjusted to the code to be printed, selectively transferring charges and creating a virtual, invisible image on the dielectric substrate.
2. This latent image is developed when passing through a magnetic brush containing black particles electrostatically opposite to those in the image (toner powder); toner deposits on the virtual image area, which now becomes visible.
3. Image is fixed to dielectric paper, blending toner by pressure and/or heat. From then on, toner is fixed to paper surface regardless of electronic balance.

h) Narrow Web Flexography

Basically similar to the system described in Section 7.A.2, this one applies to small printing machines where substrate width is usually less than 24'' (600 mm). These machines are designed for printing labels and tapes in several colors; they can be central-drum type or independent units combined with other systems.

Similar limitations to those described for traditional flexography do apply regarding bar direction, plate type and bar code contrast. Printing tests should also be performed here using the printability gauge on all machines and with all substrates. Compressible photopolymer plates are strongly recommended, engraving them directly from master films already affected to required BWR.

The picture shows a narrow web stack-type flexography printer for labels and tapes, capable of performing two-color printing at 300 m/min.

Figure 9-3

- Label printing systems chart

	Direct Thermal	Thermal Transfer	Impact	Laser	Ink Jet	Dot Matrix	Electro static	Narrow Web Flexo
High density	no	yes/no	yes/no	yes	no	no	yes	no
Medium density	yes	yes	yes	yes	yes/no	yes/no	yes	yes
Low density	yes	yes	yes	yes	yes	yes	yes	yes
Printing speed	medium	medium	low	high	high	medium	medium	high
1 label print	yes	yes	yes	yes/no	yes	no	no	no
Special paper	yes	no	no	no	no	no	yes	no
Print quality	good	excellent	good	excellent	low	low	good	medium
Unit price	low	low	low	medium	high	low	high	high
Label price	medium	medium	low	medium	low	low	medium	low
In-line labeling	yes	yes	yes	no	yes	yes/no	no	no
Symbols	most	most	one	most	some	some	most	some
Main users	in-site store	in-site store		label supplier	in-site plant	in-site office	label supplier	label supplier

Table 9-2

2. Pre-printed Pressure-sensitive labels

Most label printing systems used for bar code applications limit their action to printing in-site black bars, using the label's natural background as spaces. Background should obviously be white, but could be red, orange or yellow provided that contrast and reflectance fall within specifications (see Section 6.G).

According to the printing system chosen, care should be taken with the substrate if it is of the thermal type, as discussed in the thermal printing section. Labels should meet all bar code specifications, including left and right quiet zones, which should always be observed, and if possible, enlarged, particularly in the case of pre-printed labels.

Labels bearing stock and customer's name or "thank you" readings are very popular. They are usually printed in various colors, leaving a blank for bar code printing as well as other information like date, weight, unit price and total price.

In these cases, bar code printers are different from preprinting machines, and they are separated from one another. Care should be taken to leave enough blank space for bar code printing and for its alignment regarding other printed data.

Figure 9-4

3. Pressure Sensitive Systems for Labels

We should always keep in mind that the purpose of a pressure sensitive label is that it should stick to a given surface, staying there throughout the item's entire commercial life, without collapsing or being removed. This should be so in every circumstance, even if the item is wet, cold, hot, frozen, heated or chilled. Other factors affecting label adhesive properties are ultraviolet radiation (direct sunlight), infrared radiation (heat), and, obviously, aging of paper stock, of pressure-sensitive paper, of printed labels and the item labeled.

For these reasons, a professional manufacturer of pressure sensitive material should be contacted to indicate what the most suitable adhesive is in every case, as no such thing·as a universal adhesive currently exists. Recommendations should be passed on to users after testing every pressure-sensitive substrate available, before making a decision on label production.

Over a hundred pressure-sensitive adhesives (PSA) are available for labels and tapes. They can be basically grouped into four major categories:

a) Emulsions (water borne). For general use, medium tack and peel, removable labels (low peel), wall papers, adhesive tapes, highly cohesive, low temperature, framing films.

b) Acrylic solution (solvent based). General use, plastic film label lamination, shrink-resistant tapes, high clearness/transparency applications, high tear-out resistance, medium-high tack and peel.

c) Elastomer (rubber) solution (solvent base). General use, two-sided tapes, high-performance labels, silk-screen applications for urethane-type substrates, foam, very high tack, high peel for low-adhesive surfaces, and very high performance labels.

d) Hot-Melt. Very specific uses where high tack and peel are required and poor resistance characteristics are accepted.

Adhesive is a key factor in a label's performance and final cost. Its features will also depend on the amount of adhesive applied, which can reach the equivalent of the substrate in terms of gsm; selecting an adhesive demands maximum care, always keeping in mind a label's final application and conditions of use.

- Comparison of PSA technologies

Performance Property	Solvent-Borne		Water-Borne Emulsions	Hot Melt
	Rubber-Base	Acrylic		
Tack	Very High	Med-High	Med	High-V
Peel adh.	High	Med	Med	High
Adh. to low energy surf.	Excellent	Poor	Fair	Excellent
Shear strength	Med-High	Med-High	Med	Low-Med
Wet-out	Good	Very good	Fair-good	Poor
Resistance to:				
High temperature	Fair	Good	Fair	Poor
Plasticizers	Poor	Good	Good	Poor
Solvents	Poor	Fair	Good	Poor
Moisture	Good	Good	Poor-Fair	Good
UV light	Poor	Good	Good	Poor
Clarity	Poor	Good	Good	Poor
Aging	Good	Excellent	Excellent	Fair
Convertability	Fair	Good	Good	Fair

Table 9-3 (*Courtesy of Morton International*)

4. Labeling

Labels usually come in rolls or reels where each label is stuck to a silicon-base or high-slip substrate so that it can be easily removed. Removal can be manual, semiautomatic or automatic. Users are recommended to consult with label and label machine suppliers to access all possible choices, since labeling bar coded items should provide a solution to encoding problems rather than an additional problem.

10

Scanners

A. The light

Visible light, as we know, is a natural phenomenon provided by the sun. It reaches us either directly on to or reflected on to objects absorbing part of it which in turn reflect in different ways so we can "see" and distinguish between colors.

Since the late seventeenth century, two theories have coexisted on the nature of light. One theory maintains that light is a wave-type phenomenon, propagating as electromagnetic waves the same as radio waves, sound, television, communications or X-rays, which is correct. The other theory considers light to be a corpuscle phenomenon, consisting of millions of very small particles, known as photons, charged with light energy, moving around and propagating in a straight line, as waves, which is also correct.

Thus, classic and modern quantum physics alike come close to an interpretation of this puzzling phenomenon known as light, which, in combination with our eyes, enables us to "see".

A particular application of this principle will make it possible for us to gather and process all data stored in bar codes, shedding a special kind of light and watching them with an electronic eye called scanner. Light propagates as a sine wave identified by its frequency (number of cycles per second: c/s or Hertz) or by its wavelength (distance between the beginning and the end of each wave, or cycle), the former being inversely proportional to the latter.

Wavelength is measured in meters (m) or its multiples:

- ANGSTROM: one ten-millionth part of 1 mm
- NANOMETER (nm): one millionth part of 1 mm
- MICRON (μ): one thousandth part of 1 mm
- MILLIMETER (mm): one thousandth part of a meter

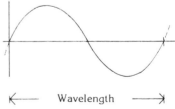

$\longleftarrow$ Wavelength $\longrightarrow$

Figure 10-1

Sunlight forms the whole spectrum of visible colors ranging from red to violet (commonly identified with the rainbow, considering that white is the aggregate of all colors and black is the absence of them all). Beyond red and violet the spectrum becomes invisible and is known as infrared or ultraviolet.

The visible spectrum covers wavelengths between 400 nm (violet) and 700 nm (red); infrared (IR) is the name given to wavelengths close to and larger than 700 nm, while wavelengths close to and smaller than 400 nm are ultraviolet (UV).

Scanning systems applied to bar codes only use a wavelength range between visible red to near infrared. It is only at these frequencies that code bar and space absorption and reflection phenomena occur within color and contrast specifications.

1. The Electromagnetic Spectrum

Red and infrared spectra are located within the range of visible wavelengths (or frequencies) shown in this chart, matched against other forms of electromagnetic radiation.

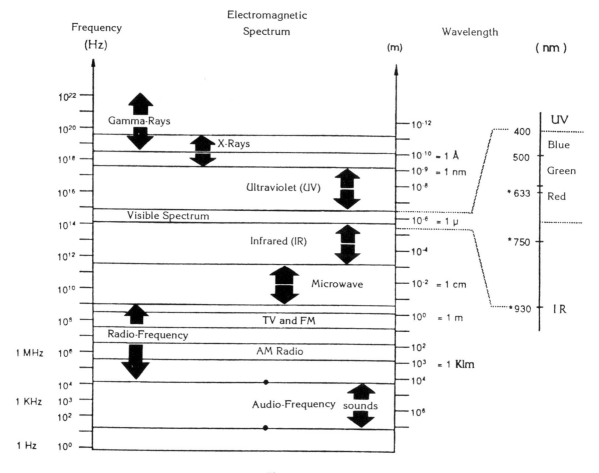

Figure 10-2

Scanners used for traditional bar codes emit and receive beam signals in three fixed, previously determined possible wavelengths that are basically as follows:

- Red (633 nm) is the most popular for commercial use at point of sale, permits the use of gas HeNe scanners. Meets all specifications described in Section 6.G. Beam is visible to the naked eye.
- Intermediate (750 nm) is slightly red, and part of its spectrum is IR; features LED capability, but light green and blue bar scanning becomes more difficult; is relatively invisible to the naked eye.
- Infrared (930 nm) scans "through" safety films and contaminants (oils, fats); require the use of high-carbon bar ink. Not suitable for organic thermal paper, totally invisible to the naked eye. Mainly for industrial application.

Wavelength selection is strongly dependent on substrate, printing, code purpose, operation and symbol destiny. Scanner manufacturers choose the beam wavelength most suitable for each specific piece of equipment; on most scanners this is not a customer option.

B. The Scanners

The name *scanner* designates the optical reading instrument capable of emitting and collecting a coherent (laser) or non coherent (LED) red, intermediate or infrared beam. Scanners explore symbols by shedding light across them and collecting the reflected beam with an optical transducer, converting a visible or invisible electromagnetic wave into an analogical electric signal.

A decoder is an electronic circuit needed to transform this signal into digital form (succession of ones and zeros in the form of pulses) and process it, checking that it corresponds 100 percent to the programmed bar code. The code is identified and transmitted to a central computer. In some cases the decoding circuit is located inside the scanner case.

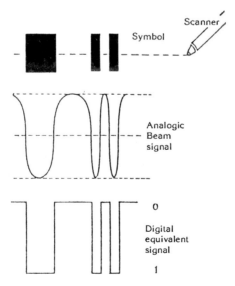

Figure 10-3 *Printed to Analogic to Digital Symbol representation*

The scanner can be used to collect data or perform symbol quality control. Selecting a suitable scanner for every use is obviously a complex task and should be entrusted to technologically qualified organizations.

The different types of most commonly used scanners are described here for illustration purposes. The basic scanner-object combination can be divided into two groups, depending on whether it is a hand-held scanner (1) or a fixed scanner (2), as shown in the scanner selection guide (Table 10-1).

Figure 10-4 *(1) Hand Held Scanner, the symbol is fixed*

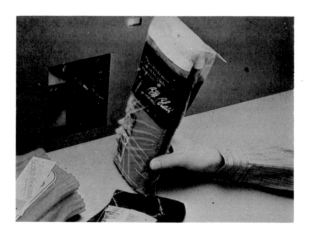

Figure 10-5 *(2) Fixed Scanner, the symbol is in motion*

- Scanner Selection Guide

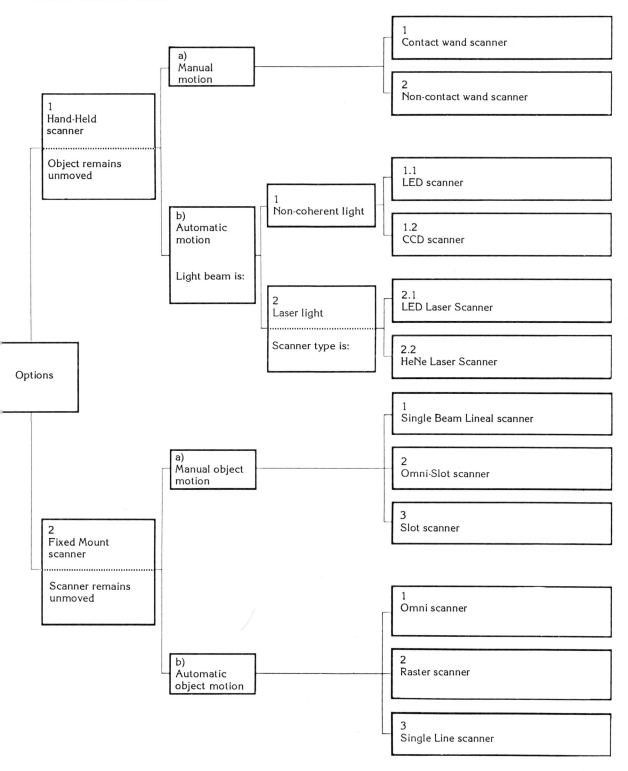

Table 10-1

Figure 10-6 *A complete family of scanners (Courtesy of Allen-Bradley)*

1. Handheld Scanner

The object remains unmoved while scanner or light motion, is manual (a), or automatic (b).

a) Manual Handheld scanner

Scanner is manually moved (manual wand); low price and low consumption make it suitable for small jobs. This scanner provides one reading at a time (about 1 reading per second), requires minimum training to keep speed and constant operating angle; and will not scan other than top-quality symbols, of any length. Opening is about 4-15 mils (0.1-0.4 mm).

   (1) Contact Wand Scanner   Equipped with LED transducers, this scanner usually pencil-shaped, containing one to four LEDs and a collecting photoelectric cell. Some of these are the cheapest and simplest scanner types. They can scan only two-dimension codes and must make direct contact with the code, using minimum energy consumption (3-10 mA, 5 VDC). Used in most handheld analyzers, because of direct contact with the code, focusing errors are minimum; scanning angle and direction are critical.

Figure 10-7

(2) Noncontact Wand Scanner   Cost is about three to six times higher than a contact wand. Code should be kept within a narrow field depth range, and three dimensions are controlled, so curved or irregular surfaces can be scanned. Small or truncated symbols are not properly scanned, and moving objects can be scanned in some cases. Also suitable for symbols in hard-to-reach locations.

b) Automatic Handheld Scanner

The object remains still and the scanner light moves automatically, through mechanical or electronic means. Can perform multiple scans per second and recurrent reading. No special training required, can read medium-quality printings.

(1) Automatic (noncoherent light) Scanner   Light spreads out in non-coherent form, i.e., in different directions rather than in-phase, and sometimes with different wavelengths, like some LEDs, incandescent bulbs, sunlight or candlelight. Energy is dispersed, so field depth ranges are narrow.

(1.1) Noncoherent Light LED Scanner   LEDs are solid state transducers; LED-equipped scanners have a limited field depth, usually lower than 6"" (15 cm), and a lower-than-4"" (10 cm) scanning width.

Depending on code density, this scanner may require adjustments, so it is recommended for scanning only one type of code.

Figure 10-8

(1.2) Noncoherent Light CCD Scanner   CCD, or capacitor coupled discharge, are virtual contact devices capable of scanning through real contact or up to 1"" (2.5 cm) distance, can scan codes on irregular or curved surfaces. Usually very light weight and small, with or without built-in decoder, features audio or visual scanning signal. Switch —or code— operated, this scanner offers different width fields, depending on code type. Gives good performance with no previous training and an average of 50 - 100 scans per second, according to brand. Sheds light on the code through a flat front red light (633 mm), collects flat beam reflected on a microcell panel where capacity discharge cells transform light into an electronic signal to be decoded.

Figure 10-9

(2) Laser Automatic Handheld Scanner The light travels as a coherent beam (laser), so all waves are equal in frequency and amplitude and are perfectly in phase. High-energy concentration occurs, and dispersion is so little that under theoretically suitable conditions the Moon could be illuminated from the Earth. For this reason, scanners can focus with maximum accuracy at different distances, enabling the obtaining of high field depth ranges on irregular or curved surfaces. Light beam moves through optical and mechanical means, so no previous training is required. May scan high- or low-density or truncated codes, requires no adjustment, and typical beam opening is 8-10 mils (0.2-0.25 mm), i.e., medium resolution.

(2.1) LED Laser Scanner Based on semiconductor solid state electronic devices known as light emitting-diodes (LED). Provides very good field depth (up to 18″ = 45 cm) and field width (up to 11″ = 28 cm).

Light and small, it usually features built-in decoder. Consumption is low: 0.5-0.75 W. Light beam used to be only infrared (750-930 nm), now available in the red spectrum (633 nm).

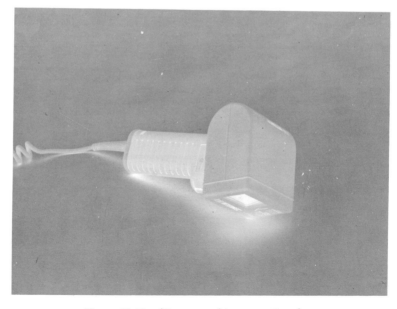

Figure 10-10 *(Courtesy of Intermec Corp.)*

(2.2) Gas HeNe Laser Scanner Transducer consists of a gas-filled tube (usually helium-neon) where high molecular excitation occurs. This generates a coherent beam of electromagnetic waves, or laser. Light is usually red (633 nm) field depth and field width are usually higher than LED lasers. Size and consumption are obviously greater than LED laser, laser tube life is about 20,000 hours.

Figure 10-11

2. Fixed Mount Scanner

This scanner remains still while the object advances manually (a) or automatically (b).

a) Manual Fixed Mount Scanner

(1) Single Beam Lineal Scanner Light beam scans visual field in a straight line, producing a visible red light that ensures accurate focusing. Models are usually compact, non bulky, and can be easily accommodated, even on desktops. Field depth is high, about 24'' (60 cm) and can operate on any code in every density.

Figure 10-12 *(Manufactured by PSC-Photographic Sciences Corp.)*

(2) Omni-slot Scanner This is the most popular scanner used at points of sale such as supermarket checkout counters. A light beam scans the visual field in three or four straight lines at the same time (omni-slot), making symbol orientation unnecessary. Usually the scanner is horizontally placed on countertops next to checkout machines, and the window usually faces up, although it can be placed vertically. Glass window should be cleaned on a regular basis and replaced annually (depending on use). Up to 15,000 items can be processed in an hour; traditional field depth is up to 10" (25 cm), and average operation rate is 500 scans per second. Scanner life is estimated at 10 years of normal use. Strong sunlight on the scanner window should be avoided as it might affect scanner behavior, mainly at high-altitude locations.

Figure 10-13

(3) Slot Scanner Scanner is low-priced and provides lower-quality performance. Items should be scanner-oriented for adequate scan.

Figure 10-14   *Slot ID-Card Scanner (Courtesy of Allen-Bradley)*

b) Automatic Fixed Mounted Scanner

Items advance automatically for this scanner, which is generally used for industrial applications.

(1) Omni Scanner This scanner provides maximum processing capacity for symbols in any position at very high speed, particularly on conveyor belts where code size and position are unpredictable. Price is high, but product ensures high productivity beyond bar code orientation (Figure 10-15).

Figure 10-15

(2) **Raster-scan** This scanner provides very high speed symbol processing at significantly lower costs but requires symbol orientation so that scanner beam can face a whole set of bars. Scanner performs up to 1 meter scanning at high speed.

Figure 10-16

(3) **Single Line Scanner** This is a medium-speed scanner for low-speed conveyor belts. The price is low when compared to other automatic fixed scanners, and this scanner provides good field depth. Symbol positioning on the item becomes critical for scanner.

Figure 10-17

Appendix A

Consulting Guide

All specific questions or comments related to this book and bar code technology can be addressed directly to the author at:

CORAS US Corp.
Attn. William Erdei
1428 Brickell Ave. Suite 300
Miami, Florida 33131 - USA
Fax: (305) 358-9014

also at:
CORAS SA. Arg.
Attn. William Erdei
Av. del Tejar 3650, Buenos Aires (1430), Argentina
Fax: (541) 544-0836
Phones: (541) 541-6656 or 541-9176

Appendix B

Figure Index

Appendix C

Table Index

Appendix D

Bibliography

AECOC, EAN Manuals
AIM, Auto ID Manuals, USS Specifications
AIM, Scan-Tech Seminars and Proceedings
Alonso M.- Finn E., Quantum and Statistical Physics
Anderson and Vreeland Inc., Flexography and Photopolymers
ANSI, Bar Code Print Quality Guideline
Bushell R. Jr, Getting Started with Bar Codes
CC1, Inc., Technical Literature
Casals R., Little Offset
EAN, General Specifications
EAN, Newsletters, Reports
Harmon C- Adams R., Reading Between the Lines
Modern Plastics, Guide to Plastics
Milprint Inc., UPC Data
Mobil Chemical International, Technical Literature
Morton International, Technical Literature
Modern Packaging, Package Engineering
Noguera E. B., Bar Codes in the Graphic Industry
RJS, Inc., Technical Literature
Schiavi-Padane, Technical Literature
Semat H., Atomic and Nuclear Physics
Stork Screens, Technical Literature
Stork Cellramic, Technical Literature
UCC, Manuals
Uniroyal, Flex-Light Technical Literature

Esta obra se terminó de
imprimir en septiembre de 1993 en
Gráficas Monte Albán, S.A. de C.V.
Fraccionamiento Industrial La Cruz
Villa del Marquez, Qro.
Apdo Postal 512, México

Se tiraron 2000 ejemplares